The Basics of
REBREATHER DIVING

Jill Heinerth

Published by:
Heinerth Productions, Inc.
5989 NE County Road 340
High Springs, Fl 32643 USA

First published 2014
Copyright © Jill Heinerth
Photography, illustrations and text
by Jill Heinerth

All rights reserved. No part of this book may be reproduced in any form or by any electronic or mechanical means, or stored in any retrieval system, or transmitted in any form by any means, electronic, mechanical, photocopying, recording or otherwise, without permission of the publisher.
All registered trademarks acknowledged.
This manual is not intended to be used as a substitute for proper dive training. Rebreather diving is a dangerous sport and training should only be conducted under the safe supervision of an active rebreather diving instructor until you are fully qualified, and then, only in conditions and circumstances which are as good or better than the conditions in which you were trained. Careful risk assessment, continuing education and skill practice may lessen your likelihood of an accident, but are never a guarantee for complete safety.
This book assumes a basic knowledge of diving technique and should be used to complement a training course specializing in rebreather diving techniques.

Cover Photo: Jill Heinerth swims in her local cave at Ginnie Springs. Photographed by Mark Long.

Back Cover: From top left: Bob Ferguson, Kim Smith and Dmitri Gorski. Photos by Jill Heinerth.

Book design by Heinerth Productions Inc.
www.IntoThePlanet.com
Printed in the U.S.A.

ISBN# 978-1-940944-00-5
CreateSpace ID 4503793

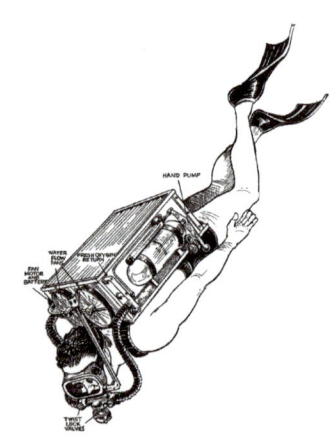

Acknowledgements

In a project of this magnitude, there are many people who deserve thanks. My deepest gratitude goes to Dr. Bill Stone who taught me about welcoming failure and recognizing it as a building block in achievement - learning when to turn back and call it a day. Sincerest thanks also go to my friend and colleague Kevin Gurr, who always believes in my abilities and kindly offers wisdom, support, ideas and friendship every step of the way. I must also thank an early instructor, Dale McKnight, who applauded me when I aborted my first deco dive, rewarding me for good judgment just when I needed that important feedback. My dear friend Kristine Rae Olmsted and her entire family have been some of my greatest supporters, encouraging me when times got challenging, and Jim Bowden, my frequent sounding board, listened to me rant and reminded me about why we do this thing called exploration.

Equipment manufacturers along the way have helped me stay in this very expensive business. I hope we all created mutually beneficial relationships that moved us each forward. VR Technology, Hollis, Suunto, Santi, Light Monkey, Waterproof, Ursuit, Aqualung, Aquatica, Dive Rite and so many others have been very generous with support through the years.

I owe a great debt to the many friends and colleagues that lost their lives, but illustrated a safer path for us to follow. They will be missed and always remembered for their contributions and dedication to the sport.

I also want to thank the multitude of dive partners, expeditionary colleagues and dear friends who either adorn my photos or overwhelm my brain with memorable experiences: Bill Adams, Matthew Addison, Sam Bennett, Scott Blair, Curt Bowen, Kenny Broad, Michael Bryant, Nick Caloyianis, Ron Carmichael, Pedro Cervantes, Craig Challen, Alex Chequer, Klaus Christiansen, Mel Clark, Christian Clark, Megan Cook, John Conway, Jo Cribley, John Dalla-Zuanna, Noel Dillon, Peter Eklund, Ali Falcone, John Falcone, Amy Ferguson, Bob Ferguson, Andrew Fock, Mike Goodheart, Brett Gonzalez, Dmitri Gorski, Kevin Gurr, Poby Han, Richard "Harry" Harris, RB Havens, Paul Heinerth, Tom Iliffe, Tom Johnson, Brian Kakuk, Carrie Kohler, Richie Kohler, Rasmus Lauritsen, Sang Jun Lee, Steve Lewis, Graham Maddocks, Jason Mallinson, Andreas "Matt" Matthes, Curt McNamee, Patti Meg Diver, Patty Mortara, Tom Mount, Gil Nolan, Kristine Rae Olmsted, Casey Omholt, Lynn Partridge, Bruce Partridge, Andrew Poole, Uwe Rath, Paul Raemakers, Jakub Rehacek, Eric Reintsma, Ben Roberts, Bruce Robertson, Jim Rozzi, Becky Kagan Schott, Phil Short, Ken Smith, Kim Smith, Peter Sotis, Eric Stadtmueller, Gregg Stanton, Bill Stone, Mitsuyoshi Tanaka, Terry Thomas, John Vanderleest, John Weisbrich and Jeff Wollenberg.

I also wish to thank the following people for additional photos in this text: Nic Alvarado, Cis Lunar Development Labs, Inc., Richard Harris, Richie Kohler, Mark Long, Richard Nordstrom, Hannah Nowill, Gene Page, US Deep Caving Team, Inc., Wes Skiles, Robert McClellan and Leon Scamahorn.

Most importantly, I wish to express the deepest gratitude to my loving husband Robert McClellan, who supports me in every aspect of my life and career. I was looking for a life partner and somehow got the amazing bonus of a talented writer, editor, new media expert, sound designer, shipping department and manager of the unmanageable. I am a very lucky girl.

ABOUT THE AUTHOR

Jill Heinerth

More people have walked on the moon, than have been to some of the places that Jill's exploration has taken her right here on the earth. From the most dangerous technical dives deep inside underwater caves, to searching for never before seen ecosystems inside giant Antarctic icebergs, to the lawless desert border area between Egypt and Libya while a civil war raged around her, Jill's curiosity and passion about our watery planet is the driving force in her life.

Jill's accolades include induction into the Explorer's Club and the inaugural class of the Women Diver's Hall of Fame. She received the Wyland ICON Award, an honor she shares with several of her underwater heroes including Jacques Cousteau, Robert Ballard and Dr. Sylvia Earle. She was named a "Living Legend" by Sport Diver Magazine and selected as SCUBA Diving Magazine's "Sea Hero of the Year 2012."

In recognition of her lifetime achievement, Jill was awarded the inaugural Sir Christopher Ondaatje Medal for Exploration. Established by the Royal Canadian Geographical Society, the medal recognizes singular achievements and the pursuit of excellence by an outstanding Canadian explorer.

With her latest project, the "We Are Water" project, she has produced a documentary film, a live presentation and interactive web resources to help steer an educational effort for everyday behavioral changes that will lead to greater access to and preservation of our endangered fresh water resources. In support of this effort, Jill and husband Robert McClellan rode their bicycles 4,300 miles across Canada, from British Columbia to Newfoundland in 2013, meeting people and through presentations to groups large and small, spreading the message of "Water Literacy." Her website, www.IntoThePlanet.com, provides links to her exploration and water advocacy efforts. Her blog www.RebreatherPro.com reaches tech divers all over all the world.

BY RICHIE KOHLER

Foreword

Since I first donned a mask and put my head underwater, I have loved diving. Like Jill I was introduced to SCUBA diving early in life, she in the Great Lakes and I in the dark cold waters of New York. Also like Jill, I would become a transplanted northerner whose diving roots took firm plant in Florida's temperate clear waters. Though the type of diving we do and the equipment we use have changed many times through those thirty plus years, the passion we each have for being underwater has never waned.

By my very nature I do not embrace change quickly. To a fault, I can be quite pragmatic, gladly leaving the pioneering of new dive equipment and techniques to others. I was definitely the "if it ain't broke, don't fix it" type of guy when it came to my diving. It has been this way for me since I first strapped on fins and it permeates every aspect of who I am. Only after a period of observing something consistently working and being advantageous to my style of diving, then maybe, I will switch over. Don't TELL ME, SHOW ME. I must observe firsthand all the pros and cons before making a fully informed and experienced choice to change either the way I was diving or the equipment I used.

But nothing stays the same, and change is inevitable. In the 1980"s it was the introduction of Nitrox and dive computers, with all the ballyhoo, controversy and teething issues that trumpeted their arrival on the dive scene. On its heels in the 1990's it would be use of the evil "voodoo" Trimix and just like with switching to Nitrox, I was at first reluctant, and then slowly adapted, opening my path to deeper and safer dives. With the new technology, equipment and techniques there were problems and accidents. Not everyone believed the rules applied to them, often with preventable but nonetheless tragic results. But with education and experience the risk becomes mitigated and these advances heralded the gold rush of new techniques and equipment, and "technical diving" came to be.

In the dawn of the new millennium it would be the closed circuit rebreather that would truly revolutionize the way I thought about, planned and conducted my diving. But again, I was loath to change, or at least to change quickly, (are we seeing a pattern here?). It took me two years of diving alongside three others on CCR before I was ready to even think about switching to a rebreather. While working on the History Channel program "Deep Sea Detectives," I watched as my team mates made longer dives, with shorter decompression times. We operated in nearly every imaginable diving environment around the world with the three of them racking up hundreds of underwater man hours. The equipment worked, and worked well. While diving on rebreathers, they were warmer and operating more efficiently than I was, and the team would often surface way before me. More than once I would surface after a dive to find they had eaten my lunch. It was time for me to take the CCR plunge.

Even after all the time I had diving side by side with rebreather divers, I was amazed at how much I needed to rethink and learn about diving once I started my formal instruction. I was humbled the first minute I strapped on a rebreather. All my years of diving experience on open circuit were of little use in mastering the operation and understanding how a rebreather functions. The classroom physics were certainly an eye opener as were the buoyancy characteristics. I came to realize that although they required a commitment of time and energy, rebreather diving rewarded me with so much more in trade.

As a diver I have always envisioned us as part astronaut/part Indiana Jones; a mix of complex equipment fused with science to the heart-racing excitement of what we will discover on the next dive. But it's more than just the thrill that keeps us hooked, in no small part it's the people- the global "tribe" of people just like us who share this wonderful experience and the identity to be a "Diver." No fiction writer could imagine a more diverse melting pot of folks that make up the core of our sport. No matter where you hail from, or what type of diving you do, one tenet has remained true since your first dive; "take care of your dive buddy, and they will take care of you." As card carrying members of this tribe, we each have an obligation to look out for our fellow divers and to share with them what we have learned in order to make the sport safer and more enjoyable.

On many levels Jill Heinerth is one of the true pioneers in our sport, and with her vast experience on rebreathers, and as a technical diving instructor, she is the perfect person to have put together this guide. On the following pages is nearly everything you could want or need to know about rebreathers. Whether you are trying to decide if a rebreather is right for you, have questions about how they work, or want to understand the differences between various types and models, the answers follow in a clear, easy to read and understand format.

For those already engaged in CCR diving Jill has peppered the book with anecdotal stories and "what if" sections, borne out of the reality of near misses and tragic accidents that help to illuminate and drive home the importance of a particular point, along with personal insights gleaned from experience that can help even the most seasoned CCR diver become more aware and a better diver. I have enjoyed and learned much in reading this guide, and no matter what your level of experience or diving aspirations may be, I am sure you will too.

Safe Diving,

Richie Kohler,

November 2013

TABLE OF CONTENTS

About the Author 4
Foreword 5
CHAPTER ONE – Introduction 11
CHAPTER TWO – The Basics 12
What is a Rebreather? 13
Rebreather History 14
Rebreather Communities 18
Rebreather Subsystems 20
   Breathing Loop 20
      Counterlungs 21 • Canister 21 • DSV/BOV 21 • Hoses 22 • Non-Return Valves 22
   Gas Pneumatics 23
      Tanks 23 • Regulators 24
   Electronics 24
      Primary Controller 25 • Secondary Computers 25 • Heads-Up Displays 25
How Rebreathers Work 26
Types of Rebreathers 27
      Semi-Closed Circuit 27 • Active-Addition 27 • Passive-Addition 27 • Semi-Closed Intelligent 28 • Closed-Circuit 28

CHAPTER THREE – Equipment Selection 31
Is Rebreather Diving Right for You? 31
How to Select the Right Rebreather 37
      Comparing Rebreather Features 40 • Comparing Warning Systems 40 • Comparing Automation Systems 41 • Comparing Mechanical Features 42 • Comparing Physical Characteristics 42
   Auxiliary Equipment 43
      Open Circuit Bailout Equipment 43 • Common Tanks 46 • Exposure Suits 50 • Writing Tools 51 • DSMB 52 • Backup Instrumentation 52 • Cutting Tools 53 • Mask, Fins, Snorkels 53 • Full Face Masks 54 • Weight Systems 55
Consumables 57
      Carbon Dioxide Absorbent 57 • Canister Design 59 • Batteries 60 • Gas 61 • Sensors 61 • Budgeting for Consumables 63
Basic Rebreather Care 63
      Cleaning Electrical Contacts 64 • Tools 64 • Spare Parts 65

CHAPTER FOUR – Training 67
How We Learn 67
Training Programs 69
      Type R Classification 70 • Type T Classification 71
Choosing a Training Agency 71

TABLE OF CONTENTS

Finding the Right Instructor 72
Preparing for Class 73
Class Content 74
Staying Current 76
    The Right Tool for the Job 77 • Specialty Environments and Training 78 • Upgrades and Modifications 77 • Specialty Equipment 80

CHAPTER FIVE – Physics 83
Calculating Pressure 83
Using Dalton's Law 84
    Converting Depth into Pressure 84 • Choosing the MOD of Bailout or Diluent 85 • Choosing the Ideal Bailout Gas 85

Rebreather Setpoint.. 85
    Floating Setpoint 86 • Choosing a Setpoint 87 • Surface Setpoint 88

Equivalent Air Depth 89

CHAPTER SIX – Physiology 91
Respiration 91
Work of Breathing 92
Oxygen Metabolism 94
    Oxygen Consumption Chart 95 • Gas Usage 96

Problems 96
    Hypoxia 97 • Hyperoxia 98 • Whole Body Toxicity 100 • Hyperoxic Myopia 102 • Middle Ear Oxygen Absorption Syndrome 102 • Hypercapnia 103 • Carbon Monoxide 107 • Chemical Injury 108 • Decompression Illness 109 • Narcosis 109

CHAPTER SEVEN – Procedures 113
Accident Analysis 113
Dive Logistics 117
    Gas Planning and Fills 117 • Calculating your SAC Rate 118 • Tank Baseline 118

Oxygen Planning 121
    Choosing the Best Diluent and Bailout Gas 121 • NOAA Oxygen Exposure Limits 122 • Residual Oxygen Toxicity 123 • Calculating Pulmonary Oxygen Toxicity 124 • Choosing the Best Mix 125

Time, Depth and Distance Planning 126
Decompression Theory and Procedures 127
    Decompression Algorithms 132 • Repetitive Dives 134

Packing the Canister 135
Pre-Dive Checks 136
    Calibration 137 • Altitude Calibrations 138 • Preventing Calibration Issues 139 • Voltage Checks, Linearity and Current Limitation 139 • Wet Sensors 141 • How Old Are Your Sensors? 142 • Pre-Breathe 142 • The Danger Zone 143

TABLE OF CONTENTS

In-Water Procedures 144
    Loop Protocols 144 • In-Water Checks 145 • Descent and Verification of Safe Operations 145 • Bottom Time 146 • Minimum Loop Volume 147 • Maintaining Buoyancy 148 • Ascents 149 • Reaching the Surface 149

Mixed Teams 151

CHAPTER EIGHT – Failure Modes 156
Trusting Technology 156
Accident Analysis 156
Common Problems and Solutions 158
    Confusing Data 158 • Diluent Flush 159 • Catastrophic Loop Failure 159 • Dewatering the Loop and Partial Flooding 160 • Carbon Dioxide Breakthrough 160 • Manual Control 162 • Gas Supply Failures 162 • Regulator Failure 162 • Mask Clearing 162 • Open Circuit Bailout 163 • SCR Mode 164 • Electronics Failures and Features 164

CHAPTER NINE – Rescue 167
Medical Emergencies 167
    Assisting an Unconscious Diver on the Surface 167 • Assisting an Unconscious Diver Underwater 168

Decompression Emergencies 168
    Omitted Decompression 168 • Decompression Illness 169 • Treating DCI 170 • Factors That May Contribute to DCI 172 • PFO 174 • Physician's Clearance 175 • An Important Investment- DAN Insurance 175

Recovery Operations 176
    Underwater Assessment 177 • Surface Assessment 178 • Witnesses 179 • Functional Assessment 179 • Hyperbaric Assessment 180

CHAPTER TEN – Testing
Testing and Validation 183
CE Testing 184
    Work of Breathing Tests 185 • Canister Duration 185 • Sensor Tracking 186 • Other Tests 187

CHAPTER ELEVEN – Rebreather Myths 189
CHAPTER TWELVE – Culture of Rebreather Diving 195
Rebreather Code of Conduct 199
Consensus Statements from Rebreather Forum 3.0 200
CHAPTER THIRTEEN – Travel 207
CHAPTER FOURTEEN – In Conclusion 213
Glossary 214
The Last Word 225

Introduction

I've been toying with the idea of writing a book about rebreathers for more than fifteen years. There were no entry level books when I got started, and the Internet sharing revolution was in its infancy. I was craving information, but it was quite hard to find. Every time I put pen to paper, diving practices, protocols and technology changed so quickly that I found my manuscript to be dated by the time it was ready for editing. New information was available and Internet forums brought forth a deluge of opinions, facts and fallacies. Start-up rebreather manufacturers popped on to the market and, just as quickly, disappeared. The instructor and expert pool deepened and some of the most experienced ones drowned. Several times I shelved this manuscript thinking that one day things would settle down and the technical diving and rebreather industry would reach a certain level of stability. This past year, as I rode my bicycle across Canada, I realized that this was precisely the reason I needed to write this book. There is suddenly a significant and expanding interest in rebreathers, but good information is still hard to find and separate from the rabid (and sometimes misinformed) opinions expressed on Internet forums. Most people who have purchased a rebreather for $12,000 are naturally going to think they bought the finest diving technology available. Likewise, after building a relationship with a rebreather expert, most would proclaim their instructor as the highest authority- simply the best.

This book is not for everyone. It is written for the seeker. It is provided as a "non-denominational" guide to help you understand the technology, figure out whether the risk is worth taking, and consider how to make the right investment in equipment and training. I want this book to be a conversation between us- about everything you need to know to make good decisions. I want to answer everything that you might consider to be a stupid question. I've owned, maintained, tested, and helped develop many rebreathers in the past and have come to know that there is not a perfect answer for every diver. This guide is intended to provide you with a foundation to thoughtfully examine the available information, and learn the basics about rebreather diving.

2

The Basics

In this chapter:

- *What is a Rebreather?*
- *Rebreather History*
- *Rebreather Communities*
- *Rebreather Subsystems*
- *How Rebreathers Work*
- *Types of Rebreathers*

When I was a kid, I wanted to be an astronaut. In kindergarten class, we were herded into the school library to gather around an old black and white television set. Through the hissing static, I could see Alan Shepard making a chip shot on the lunar surface and driving what appeared to be a bouncy golf cart. The astronauts were wearing space suits and were playing like excited kids. That evening, I stared at the big moon in the sky, pondering how they could be way up there…and breathing. I *really* wanted to be an astronaut.

If you are like me, then you've probably looked at rebreather divers in much the same way. Diving is your passion, but being able to descend silently into the blue depths and enter the domain of mysterious ocean creatures seems like a dream. Without bubbles, you can be a silent partner of the underwater community, not just a clamorous observer that frightens everything in its wake. If you have the mind of an explorer, then you are probably looking at rebreathers as a chance to do what the astronauts did and "go where no man has gone before." Whether it is a cave, wreck or the uncharted depths of an ocean wall, there is some purpose that this rebreather might fill in your life beyond open circuit SCUBA.

So let's start with the basics. How is a rebreather different from what you already know as a diver and what purpose might it fill in your underwater adventures?

CHAPTER 2 - THE BASICS

What is a Rebreather?

I'm assuming that you are already an open circuit SCUBA diver of some level, although these days we are getting to the point where people will be able to learn the basics of diving using a rebreather. If you think that's crazy, then you should meet Sharon Readey. She helped engineer the original PRISM Topaz rebreather, and yet, has never taken a breath from traditional SCUBA. She can't even figure out *why* someone would want to make bubbles underwater.

Open circuit SCUBA is inefficient and time limited. Much of every exhaled breath from a SCUBA tank is wasted. Our bodies only use a very small portion of the oxygen component of air or nitrox. We metabolize a tiny bit of the available oxygen and expend the rest into the water column in a stream of bubbles. A rebreather is designed to recapture all or part of that exhaled breath and recycle as much as possible for the diver to use again. As a result, we are offered many advantages over traditional SCUBA, such as a longer bottom time per unit of gas, reduced bubbling that might frighten away marine life, as well as many other positive features we'll cover in detail later in this book.

Rebreathers offer a whole new world of opportunities to extend range, interact with wildlife and gain confidence in exciting new technologies.

It sounds like a simple enough concept, but rebreathers perform some other important tasks. Each exhaled breath contains carbon dioxide, which can rapidly build up and cause our bodies distress. The rebreather has to "clean" the gas and remove carbon dioxide before sending it back to the diver. It also has to make up the lost volume of oxygen that was metabolized by the diver. It's a balancing act. The rebreather has to predictably and repeatedly add small amounts of oxygen and remove carbon dioxide to keep the life support environment optimized and safe for a diver to breathe.

An astronaut wears a rebreather on a space walk, but at every moment there is a team of experts at Mission Control watching his every move. A Medical Officer or Safety Supervisor can intervene if there is any indication of a problem. Rebreather divers are oftentimes using life support systems that are more complicated than those used by astronauts. Divers generally have a smaller safety net and less redundancy. There is no question that manipulating a life support environment is the most dangerous and sophisticated thing a person can take on. Worse yet, human beings do not follow rules very well. By reading this book thoughtfully, I will help you decide whether this risk is warranted, and if it is, how you can best mitigate the risks caused by human error.

CHAPTER 2 - THE BASICS

Rebreather History

Rebreathers have been around far longer than traditional SCUBA. If you look as far back as 400 BC, you can find artistic references to swimmers exploring below the surface while breathing on goat stomach lined bags. (Yuck!) Whether based on fact or fiction, man's imagination has taken him to the depths of the ocean for many hundreds if not thousands of years. Alexander the Great and Jules Verne come to mind.

In the more recent past, our desire to swim underwater was more a product of practicalities. As early as the seventeenth century, rudimentary breathing systems were built into crude submarines. By the early nineteenth century, experimentation began with primitive diver units. Submarine escape, bridge construction, and naval warfare fueled a race to improve the technology. Sending covert swimmers into combat allowed for stealthy operations and an element of surprise. Most large navies began their own development programs and mass production was finally a reality.

This early advertisement for the Electrolung rebreather boasted operating costs as low as $1.50 per hour regardless of depth. Photographers were targeted as customers, but the device was never fully accepted by the diving community.

A few war-surplus units landed in the hands of civilian divers, but rebreather diving mostly remained in the military and commercial fields until the late 1960s. Scientists Walter Starck and John Kanwisher met on a research vessel while using surface-supplied helmets to conduct lock-out dives. This equipment used a lot of gas and forced them to be tethered to the vessel. It was severely limiting their work. Cooperating to find a solution, the two men subsequently developed the first viable electronically controlled rebreather. The Electrolung was a true homebuilt device, sporting a valve salvaged from a cuckoo clock and pieces of an old

CHAPTER 2 - THE BASICS

breadboard. After selling the Electrolung design to Beckman Instruments, an advertisement appeared in Skin Diver magazine in 1970, appealing to avid sport divers and underwater photographers to consider using this revolutionary new tool called a "rebreather."

For years the technology remained in the hands of a few photographers, commercial divers and military personnel, but it did not inspire the confidence or imagination of the average diver. An Electrolung rebreather was priced at $2,975.00 in 1970 (about the same as an average new car). A series of fatalities resulted in Beckman Instruments withdrawing their commercial interest in rebreathers. It wasn't until the 1990's when several high profile technical diving expeditions attracted the attention of the media, that this radical diving technology broke through and began to strike a chord among the more adventurous SCUBA divers of the time. News stories proclaimed the extended range offered by rebreathers. Rebreather diving paralleled the accelerating achievements of technical open circuit SCUBA divers. Each discipline enjoyed a period of remarkable innovation and growth as advanced technology and a soaring human spirit extended the frontiers of underwater exploration and discovery.

When you sign up for a rebreather class, you'll learn a lot more about the historic landmarks in rebreather diving history, but what is important to understand now, is that this sport is still in its early stages. Recognizing we are still at the beginning of sport rebreather diving will remind you be discerning about some things you may read online. Rebreather technology and best practices are evolving rapidly and it is critical to stay current and keep up with innovation.

My first exposure to rebreathers came in late 1994 when I met Dr. Bill Stone. A genial engineer, the lanky Dr. Stone invited me into his garage. I was honored, and almost speechless to be given access to his inner sanctum of muddy toys. He had recently returned from an expedition that utilized his Cis-Lunar Mk-4R rebreathers that enabled divers to penetrate deeper into the planet than anyone had gone before. To execute the first dive required over two miles of rope to reach the water. It took dozens of cavers and weeks of time to stage base camps inside the remote mountains of Central Mexico. Using traditional SCUBA tanks was simply not feasible. An empty tank relayed back

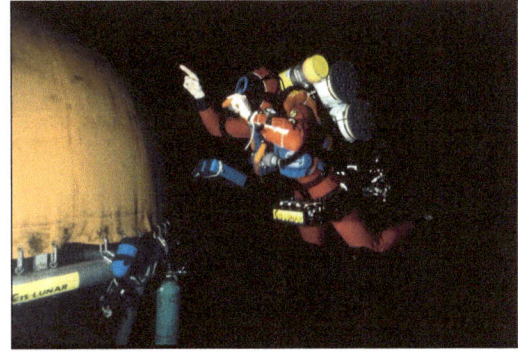

Dr. Bill Stone makes a landmark 24-hour dive on his Cis-Lunar MK1 rebreather during exploration of Wakulla Springs by the United States Deep Caving Team in 1987. His successful proof of concept launched a journey into rebreather engineering that continues to this day. Photo: United States Deep Caving Team, Inc.

CHAPTER 2 - THE BASICS

to the surface for fills would require a round trip of several days. The first dive was made in relative luxury compared to the series of remote outposts that lay beyond multiple flooded sumps and cascading waterfalls. Aided by dozens of others, he and exploration partner Dr. Barbara AmEnde were ultimately successful in using these rebreathers to reach one of the loneliest places on planet earth. James Tabor, in his book *"Blind Descent,"* would later describe it like this:

Dr. Bill Stone diving in Huautla in 1994, on the Cis-Lunar MK-4R rebreather. Photo: US Deep Caving Team, Inc.

"AmEnde and Stone's six-day penetration of Huautla, beyond all hope of rescue, ranked as one of the century's greatest, and least heralded, exploration achievements."

Stone had designed and developed each piece of the crucial life support system, and deployed it effectively, proving that inner earth exploration was no different from that of outer space.

I was already familiar with Dr. Stone's exploration and design efforts. I had read his book on the 1987 Wakulla Project and soaked up every page of information on mixed gas and staged cave exploration. Many divers of the time used this as their primer for trimix diving with its real-world examples of what was previously considered impossible. At Wakulla Springs, Stone made the first 24-hour dive on his Cis-Lunar Mk1 closed circuit rebreather. The name Cis-Lunar originated from a concept recognized in the aerospace industry. Dr. Stone was employed by the National Institute of Standards and Technology (NIST), working on a project that promised to boost the space shuttle's external tank into orbit and convert it to an industrial workshop. Working with mathematical calculations daily, he became aware of the term "cis-lunar" which was used to describe the place in space between the earth and moon. It was within this volume of space that the first commercial manned space efforts were likely to take place. Considering that the development of his rebreather was really intended for lunar exploration, this seemed like the right name for his new company, Cis-Lunar Development Laboratories, Inc. The

The author pilots the Digital Wall Mapper during the Wakulla2 Project in 1999. Photo: U.S. Deep Caving Team, Inc.

CHAPTER 2 - THE BASICS

MK1 was viewed as a prototype for space suit life support. The landmark proof of concept at Wakulla Springs was not only a success for Stone's new company, but it also launched a new era of diving technology for the masses; a virtual "space race" for the exploration of the aquatic world.

Kevin Gurr diving on a treasure salvage project in 1998. His team conducted 9,360 in-water man-hours using rebreathers to recover artifacts from a sunken galleon over five years propelling him into designing his own rebreathers.

In the late 1990s, I had the opportunity to work with Dr. Stone at the United States Deep Caving Team's Wakulla2 Project.[1] By this time I had explored challenging systems with him in Mexico, and developed a deep friendship and respect for his efforts. Stone was a man caught between two worlds. He developed specialized equipment for space exploration- and designed tools that helped him explore inner earth. His deep cave explorations were the proving ground for his dream of visiting other bodies beyond our planet. His brilliance often went right over people's heads. He was bold and confident about taking on unfathomable challenges. Yet, he was also very cautious. He had lost several friends to diving and caving accidents and recognized that the rebreather revolution would come with casualties. It was his goal to try to engineer them out.

By the turn of the new millennium, many manufacturers bravely jumped into the game. Martin Parker of AP Valves launched the Inspiration just as Stone was beginning to gain a foothold with the Cis-Lunar MK-5P. Others were prototyping and releasing new units as quickly as they could. Dräger attempted to capture the recreational end of the market with units that operated on a simpler, less expensive concept, and homebuilders were outwardly looking for eager guinea pigs in rebreather divers who were willing to subject themselves to human testing.

This very recent history was not without occasional incidents and accidents. Clearly, engineering could not

Martin Parker of AP Valves should be credited with the vision of figuring out how to open the rebreather market to the masses. Parker managed to successfully navigate through the minefield of challenges and turn a blossoming idea into a real technical diving marketplace that served its customers globally.

CHAPTER 2 - THE BASICS

fully compensate for the myriad of potential errors made by a human diver. Evidently, we knew little about how time affected the behavior of basic consumable products like sorb and sensors. There was, unquestionably, much to learn about training techniques, human nature and the effects of creeping complacency. We were learning quickly, but not fast enough to reduce the staggering rate of accidents. By the uninitiated, rebreathers were labeled as dangerous "boxes of death," while more experienced divers arrogantly boasted why "it could not happen to them."

Today, rebreather diving is associated with a significantly greater risk of fatality than open circuit technical diving. The best compilation of information I have found on accidents comes from Dr. Andrew Fock of Australia.[2] Though the database is still small and statistically difficult to analyze, it appears that users of rebreathers may face up to a ten-fold greater chance of dying, than their open circuit technical diving counterparts. With recreational units only just beginning to penetrate the marketplace, there is not yet a reasonable way to calculate potential risk for those types of divers. One might presume that because recreational sport rebreathers permit less user-control over the life support formula, they may offer a great safety margin over their more technical cousins. Only time will tell. Dr. Fock's analysis confirms what most of us already knew within the industry. It is not the the rebreathers that are killing people, it is the way that they are being misused. I feel that most rebreather accidents could have been avoided with strict adherence to accepted safety protocols. I'll be covering more on this this throughout the book.

Brett Gonzalez assists Brian Kakuk as he prepares for a deep dive on Challenger Bank on the NOAA Deep Water Caves project in Bermuda in 2011.

Rebreather Communities

Today, there are several different communities regularly employing rebreather technology for different purposes. Scientific, public service, military, commercial and sport diving markets all have different needs and issues motivating their acceptance of rebreathers.

The scientific community is slowly embracing rebreathers that facilitate the luxury of long bottom times in an undisturbed water column. There is little question that the rebreather is a remarkable

CHAPTER 2 - THE BASICS

tool for scientific discovery. Organizations such as the American Academy of Underwater Sciences (AAUS), National Oceanic and Atmospheric Administration (NOAA) as well as universities and colleges are developing tight standards for the use of closed-circuit technology on their projects. Their acceptance is slow and standards revisions are often the result of reactionary policies developed after accidents. Best described as a skittish relationship, scientists recognize the potential of the equipment while administrators are concerned about risk and liability.

Very simple oxygen rebreathers may not be state-of-the-art, but they serve a great purpose for lean, covert, military operations. Photo: Richard Nordstrom.

The military use of rebreathers has not slowed since their first regular deployments in World War I. Covert operations, submarine escape, and silent diving are more important than ever before. Yet, it is the commercial manufacturers that drive development and innovation. The U.S. and other naval forces around the world have found it more effective to team up with rapidly evolving commercial firms than to develop their own proprietary machines. Similarly, public service divers are using technology developed in the private sector. The National Park Service, police divers and rescue teams have followed the lead of active technical divers, and are using commercially developed equipment. Like the scientific diving programs, they generally operate within the oversight of government bureaucracies.

The term "sport diving" is used to encompass both recreational and technical diving pursuits, for anyone that is not operating in the area of commercial diving.

Commercial diving operators have been using rebreathers since the mid 1960s - mostly out of the line of sight of the sport diving industry. Commercial divers in the U.S. fall under guidelines developed by the Occupational Safety and Health Administration (OSHA), the federal agency charged with the enforcement of safety and health legislation, especially pertaining to individuals in the workplace. Accident analysis in commercial operations is rarely public and technology is often proprietary. Rebreather systems are supplemented with surface support and bear little resemblance to the types of autonomous devices utilized in the sport diving market.

CHAPTER 2 - THE BASICS

The term "sport diver" describes anyone that uses a rebreather for non-commercial purposes and encompasses both the recreational and technical areas of diving. This term broadly describes divers who are not currently subjected to OSHA or other governmental regulations and guidelines.

Rebreather Subsystems

There are dozens of rebreather manufacturers building different types of rebreathers around the world. You can almost compare the rebreather marketplace to buying a car. (I've owned a few rebreathers that certainly cost more than the well-used vehicle sitting in my driveway.) There are dozens of car companies that make vehicles ranging from small Smart Cars to big 4x4 trucks. Some cars run on gasoline and others employ hybrid technology. There are even human-powered vehicles such as bicycles which fall under the general category of transportation. Rebreathers can be as different as these modes of transportation, as can their costs. Regardless of their differences, cars have many similarities and contain common subsystems. Each car, motorcycle or truck has wheels. Each vehicle has some sort of engine and brakes. Some cars have airbags, anti-lock brakes and sophisticated locking mechanisms. Each vehicle has a specific purpose. Rebreathers can be broken down into similar subsystems and mechanisms. Understanding the basics about how these work, will help you make choices about the features in rebreathers that are important for the type of diving you do. Understanding subsystems and features will help you find the right tool for the job.

Breathing Loop

Every rebreather contains a breathing loop. Within this subsystem, the diver inhales fresh gas and exhales waste gas. Gas is captured in the loop and one-way valves ensure that it travels in the correct direction. The gas routes through hoses to a canister containing a filter/scrubber, which contains a material used for removing carbon dioxide. The breathing loop includes one or two counterlungs, which are flexible bags enabling changes in volume as the diver breathes. The diver acts as the engine that forces gas around the circuit.

A Megalodon CCR prepared for diving inside Atlantida Tunnel in Lanzarote. The over-the-shoulder (OTS) counterlungs are constructed of a durable Cordura fabric.

CHAPTER 2 - THE BASICS

You will find various terminology applied to similar parts across different brands.

Counterlungs/Breathing Bags

A collapsible chamber or flexible bag of some sort, connected to a rebreather breathing loop, which expands as a diver exhales, and collapses as a diver inhales. These counterlungs are generally labelled as exhalation and inhalation counterlungs. Several styles are available and are named for the location on the diver, such as: back-mounted, front-mounted and over-the-shoulder (OTS) counterlungs. Rebreathers have one or two counterlungs.

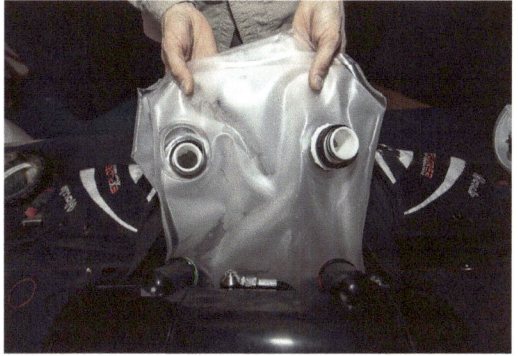

Nested, translucent inhalation and exhalation counterlungs inside the Explorer rebreather are protected by a hard case.

Canister/Scrubber/Stack/Filter

A rigid container, connected to the breathing loop, which holds the carbon dioxide absorbent material. Scrubbers may be designed in several different flow patterns and may be housed inside a canister.

Mouthpiece Block/Diver Surface Valve (DSV)/Bailout Valve (BOV)

The component of a rebreather breathing loop through which the diver breathes. This usually includes a two-way switch or lever to prevent water from entering the breathing loop and sometimes includes an integrated open circuit regulator as in the case of a bailout valve (BOV). This is generally where the critical non-return valves are installed. In some rebreathers, non-return valves are installed on the ends of the breathing loop hoses. In both cases they assure the gas travels in the correct direction.

A diver filling their scrubber canister (above). A bailout valve (below) equipped with a lever that permits the diver to switch from closed circuit operation to open circuit.

CHAPTER 2 - THE BASICS

Hoses

Breathing loop hoses direct gas from the mouthpiece block towards the scrubber-filled canister and from the canister towards the mouthpiece. These hoses are usually easily removable for sanitization.

One-way Valves/Mushroom Valves/Non-return Valves/Check Valves

These mushroom-type valves prevent subsequent re-inhalation of used gas, and promote circulation of exhaled gas towards the carbon dioxide scrubber canister. A rebreather generally contains an upstream and a downstream check valve. They are similar in appearance to the exhaust valve found in a SCUBA regulator's second stage.

> **To BOV or Not to BOV**
>
> *BOVs can be handy devices, allowing you to leave the DSV in your mouth if you need to switch to open circuit. Keep the following in mind:*
>
> *1. BOVs are designed to allow for an easy to switch to open circuit, but don't be hasty and switch back until you are certain you know what is going on in the breathing loop. Sanity "breaths" were once taught as a solution intended to clear your mind, but we now know that mere breaths will not solve an oxygen or carbon dioxide*  *issue in the body. One or two breaths will not fix a hypoxia problem if you are already feeling fuzzy. You need to take your time. If a problem is solvable and it is safe to do so, then there may be ways to switch back to the loop on some rebreathers. However, if you switched because you felt funny, it's smarter to stay off the loop and get your brain back in gear again.*
>
> *2. Your BOV must be plumbed to a meaningful gas supply if you plan to use your BOV for more than a couple of breaths. If your BOV is hooked up to a small onboard tank in the rebreather, you will have to switch to your offboard bailout tank unless you can make a short, direct ascent.*
>
> *3. You must test your BOV underwater every time you dive to ensure its function. This is done at the beginning of every dive.*
>
> *4. You should equip your offboard bailout tank with a second stage for other divers or for BOV failures. The only exception is for shallow recrea-*

CHAPTER 2 - THE BASICS

tional divers above 18m/60 ft. who don't have a bailout tank. Their second stage will be plumbed to the onboard tank and carried like an octopus.

5. If you are conducting deep trimix dives, the onboard tank may contain a hypoxic trimix that renders the BOV unsafe to use in shallow depths.

6. Your BOV should be tested and proven to be "Class A" CE rated EN250 and designed for the depths and workload you intend on using it. Any sliding valves, connectors and convenience items that are added to your BOV system must pass similar scrutiny. Anything that might restrict gas flow may increase the breathing effort which can be problematic if you are suffering from over exertion or carbon dioxide issues.

All recreational rebreathers have BOVs, but if you are learning to dive a technical rebreather, this may be an optional item. I suggest mapping out a risk/benefit list and checking all the variables so you understand the intention, advantages and disadvantages of using a BOV.

Gas Pneumatics

Rebreathers are equipped with a gas delivery system that contains components that will be very familiar to open circuit divers. At least one tank and regulator/demand valve are required to supply gas for the breathing loop.

Tanks/Cylinders

At least one tank acts as the onboard supply of gas. Many rebreathers employ two separate tanks, in this case filled with oxygen and a diluting gas. The tank that carries the diluting gas is called the Diluent Tank or Onboard Diluent. Tanks vary in size, but rebreathers generally make use of very small tanks since the gas supply is significantly extended by the technology. Two- and three-liter tanks are very common in rebreathers that use two tanks. Four- or six-liter tanks are common in rebreathers that use a single tank for their gas supply.

Both aluminum and steel tanks with screw-in DIN valves are commonly used in rebreathers. The decision to select either steel or aluminum may be based on cost, weight, buoyancy and availability.

Open circuit bailout tanks are commonly used with rebreathers, and are considered mandatory when diving below 60 feet/20 meters. These tanks are fitted with a SCUBA regulator (first and second stage), pressure gauge and sometimes

CHAPTER 2 - THE BASICS

an inflator hose and are filled with a gas that is safe to breathe at maximum depth. Theses tanks are intended as a backup if a failure demands aborting the dive. The size of the bailout tank should be large enough to allow the diver to reach the surface with a safe margin and may include consideration for escaping a complex overhead environment and buddy rescue.

Regulators/Demand Valves

Regulators are installed on the onboard tank(s) to drop the high pressure to a usable level. These regulators/demand valves may appear identical to open circuit regulators, but they may contain additional unseen technology such as flow restrictors. They are also adjusted differently than open circuit regulators and should only be serviced by a certified technician familiar with the proper specifications. The oxygen regulator must be cleaned and designed for oxygen service. The diluent regulator may need oxygen service depending on local regulations. In some parts of the world, the oxygen tank valve and oxygen regulator DIN fitting are a different size than North American versions. Check this out before traveling overseas.

Regulators may be equipped with hoses that specifically fit the particular model of rebreather. These hoses may also be constructed of a more flexible material than standard rubber hoses in order to fit inside of a rebreather case. Low pressure hoses supply the breathing loop, inflator mechanisms and bailout valves or second stages. High pressure hoses will be attached to standard submersible pressure gauges that show tank contents. Some regulators are fitted wireless "senders" that transmit pressure information to the diver's handset.

Electronics

Many rebreathers, but not all, are electronically controlled. Electronic rebreathers are referred to as eCCRs or eSCRs. The function of the electronics package is to add oxygen to the system as needed and warn the diver of developing problems through visual, audio and/or tactile alarm systems. The expense of a particular rebreather may be directly proportional to the features offered in the electronic systems.

CHAPTER 2 - THE BASICS

Primary Controller

Many rebreathers have a primary controller which acts as the brain of the unit. Depending on the unit purchased, it may perform tasks such as controlling an electromechanical solenoid valve, interpreting data from oxygen and/or carbon dioxide sensors, logging dive parameters and issuing alarms for arising problems. Current design trends are focussed on providing more "black box" capacity that gives a diver or manufacturer downloadable detailed information from actual dives and firmware upgrade capability.

Primary Display

Essential safety information is displayed on a primary handset which is generally worn on the wrist. The diver checks this display every one to four minutes to confirm life support function and levels and other critical information such as remaining consumable levels and decompression status.

Cave diver Jeff Wollenberg has placed his DIVA to be easily viewed through the lower corner of his mask. The vibrating motor will be felt through the DSV.

Secondary Computers

Many eCCRs are equipped with a secondary display which may be completely independent from the primary display. These are most often offered as handsets or consoles.

Heads Up Display/HUD/DIVA

A Heads Up Display (HUD) is a type of secondary display which gives the diver information about critical life support functions. These displays are generally armed with a series of lights whose color, position and flashing status inform the diver about oxygen levels and sometimes other functions such as ascent rate, decompression status and developing problems in the unit. Some are equipped with a vibrating motor that acts as a tactile alarm. This style is often referred to as a DIVA (Display Integrated Vibrating Alarm).

CHAPTER 2 - THE BASICS

Sensors/Cells

Rebreathers use galvanic sensors to detect the level of oxygen within the breathing loop. Electrical current from the cell(s) is interpreted by the primary controller to inform the diver about the partial pressure of oxygen in the loop.

Some rebreathers are equipped with carbon dioxide sensors that inform the diver of dangerous situations where carbon dioxide levels may be rising above a safe threshold.

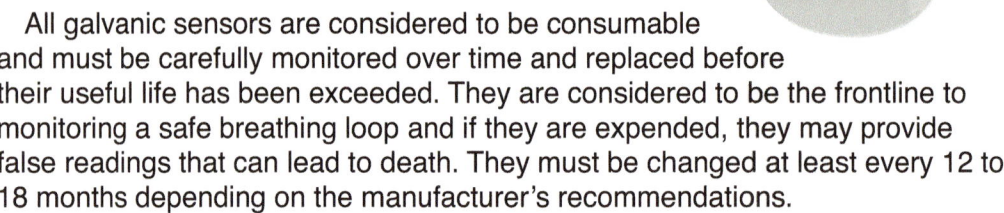

All galvanic sensors are considered to be consumable and must be carefully monitored over time and replaced before their useful life has been exceeded. They are considered to be the frontline to monitoring a safe breathing loop and if they are expended, they may provide false readings that can lead to death. They must be changed at least every 12 to 18 months depending on the manufacturer's recommendations.

How Rebreathers Work

In general, rebreathers are designed to recycle the gas that a diver normally exhales into the water column. This action conserves gas volume, reduces the noise made by bubbles and generally extends the range of a given dive. As with automobiles, rebreathers may be operated manually or automatically are made by numerous manufacturers that offer different features and designs.

A diver exhales into the breathing loop. Thanks to one-way mushroom valves, the gas travels in one direction only. The breathing loop is equipped with one or two flexible counterlungs that help compensate for the shifting gas volumes that occur through the respiratory cycle. Gas is routed to the scrubber which contains a carbon dioxide absorbent material which removes the carbon dioxide from the exhaled breath. After dwelling in the absorbent material, the gas passes over one or more oxygen sensors. If the oxygen level warrants, oxygen may be added with an injection or flow of oxygen or other high PO_2 gas, bringing the life support levels back to a suitable point. The gas may pass through an-

CHAPTER 2 - THE BASICS

other flexible inhalation counterlung and then through a one-way valve to the diver's mouth. Sensors carefully monitor oxygen levels. Either the diver and/or an electronics package intervene and add oxygen as needed.

Types of Rebreathers

Semi Closed Circuit Rebreathers (SCR)

Semi Closed Circuit Rebreathers are generally simple in form and function. In this type of rebreather, only one tank of gas is provided onboard. Depending on the design, the gas may enter the breathing loop in a number of different ways. It may flow continuously, it may be injected based on the breathing rate of the diver or may be added intelligently by an electronic system as required.

When a diver's exertion rate is high and when a low percentage of oxygen is carried in the tank, then a greater flow of gas must be provided in order to keep oxygen partial pressure at an acceptable level. With more inflowing gas, the diver or unit may vent bubbles at a greater rate, resulting in faster gas use. If a higher percentage of oxygen is carried in the onboard tank or if the exertion of the diver is lower, then venting of bubbles may be reduced and the gas may be used at a slower rate in some models. In SCRs, depth should never exceed the safe partial pressure of the gas in the supply tank.

SCRs are not common in the technical end of rebreather diving because they do not offer the decompression advantage that fully closed circuit rebreathers present for a diver.

The method of gas addition helps to describe different models of SCRs:

Active-addition – a rebreather gas-addition system that actively injects a continuous stream of gas into the breathing loop (such as a constant-mass flow valve in certain kinds of semi-closed rebreathers). Imagine breathing in and out of a paper bag. If a stream of continuous fresh gas is fed in to that paper bag, then some will be inhaled and some will bleed out around the edges. If you breathe fast and metabolize more oxygen molecules, there may be less gas bleeding out of the bag. If you breathe slowly, then more gas will be wasted and vented.

Passive-addition – a gas addition system utilized by some SCRs to passively inject gas into the breathing loop; usually achieved by a mechanical valve that opens in response to a collapsed bellow or drop in breathing

The Hollis Explorer eSCR is designed with intelligent addition of gas to maintain a floating setpoint. This new class of rebreather meets the needs of the recreational market.

CHAPTER 2 - THE BASICS

loop gas pressure. In this case, imagine that as you breathe in and out of the paper bag, you slowly deplete the volume due to metabolizing the oxygen in the bag. When the volume drops in the bag, it gets harder to breathe. Passive addition valves will be triggered to flow with a drop in volume in the bag.

Electronically Monitored mSCR – a mechanical SCR with electronic monitoring. In this case, electronics are used to inform the diver of PO_2 as well as provide warnings and status updates, however the gas control is manually controlled by the diver.

Semi-Closed Intelligent – an electronic gas addition system utilized by the newest class of eSCRs to intelligently inject gas into the breathing loop when a drop in partial pressure warrants an addition. In this way they can match changing metabolism better than their mechanical counterparts. In this example, an oxygen monitoring and control device has been added to the system. When the PO_2 drops, fresh gas is added to maintain a suitable floating setpoint (desired PO_2 in the breathing loop) to optimize gas use and other consumables and to adjust for changing exertion (oxygen consumption) rates of the diver. Gas also flows into the bag automatically if a pressure drop warrants. A failsafe mode reverts the eSCR to the active addition mode in the event of total failure.

By their nature, semi-closed circuit rebreathers make more bubbles than closed-circuit rebreathers. They are designed to vent in order to make room for higher partial pressure gas that brings the PO_2 up. Active-addition style SCRs make more bubbles than either passive or intelligent styles and all rebreathers will vent bubbles upon ascent. A valve called an Over Pressurization Valve (OPV) is fitted onto all rebreathers. Whenever the volume of gas in the breathing loop exceeds a given level, such as upon ascent, then bubbles will vent through the OPV.

Closed Circuit Rebreathers/CCR/ Fully Closed Circuit Rebreathers

CCRs are the most popular type of unit for technical divers. They offer the greatest decompression advantage since they create a virtual mixing station on your back. As a result, they have their own inherent risks.

In this type of rebreather, two bottles of gas are supplied onboard. One tank is filled with pure oxygen and the other tank is filled with diluent gas. When you start diving on this type of rebreather, the diluent tank will likely be filled with air. As you descend in the water column and the surrounding pressure increases, the flexible counterlungs start to collapse much like the air cell in your BCD. As it

CHAPTER 2 - THE BASICS

collapses, an automatic diluent valve will open and diluent gas will flow into the breathing loop. While descending, the partial pressure of oxygen naturally rises from increased pressure and you are also metabolizing oxygen. You may reach your target depth before you need more oxygen molecules in the loop.

When the partial pressure of oxygen drops, then a valve is triggered either manually by the diver or automatically through an electronics system. A tiny addition of oxygen brings the breathing mix back up to a safe PO_2. Generally, the oxygen injection occurs in a location ahead of the the scrubber. In the scrubber, the gas mixes and homogenizes and is cleansed of carbon dioxide. When it exits the scrubber, it passes by one or more oxygen sensors. The PO_2 is displayed in the computer handset and triggers the primary controller (in an electronic rebreather) to add oxygen as required.

As your dive commences and you metabolize oxygen, more oxygen will be needed to make up the balance. Further injections of oxygen will be triggered by the diver or the electronics depending on the type of rebreather.

The author diving her PRISM2 rebreather. Photo: Mark Long.

In a CCR, oxygen is added judiciously and as a result, there is less likelihood of having too much gas in the breathing loop causing a situation where gas is vented as bubbles into the water column (if you remain at a stable depth). If the rebreather is piloted well and the diver has a good mask seal, then bubbles will only vent on ascent (including small rises in the dive profile) as the gas volume in the breathing loop expands.

The act of adding oxygen in CCRs happens in a number of ways. In a manual rebreather, the diver closely watches their handset, which informs them about the PO_2 inside the breathing loop. When PO_2 drops below their target setpoint, the diver pushes a button that causes oxygen to flow into the loop. Careful and constant monitoring is required. This type of rebreather is also referred to as a Diver Controlled CCR or dcCCR. Electronic CCRs allow the diver to choose a PO_2 setpoint. The onboard computer monitors PO_2 and adds oxygen when the PO_2 drops below the chosen setpoint. Most eCCRs are equipped with a manual addition valve and can function just like a diver controlled CCR if desired.

1. http://www.stoneaerospace.com/about-us/about-us-history.php
2. Fock, Andrew W. (2013), Analysis of recreational closed-circuit rebreather deaths 1998-2010, Diving and Hyperbaric Medicine, Volume 43, No. 2, June 2013, pp 78-85 http://www.dhmjournal.com/files/Fock-Rebreather_deaths.pdf

3

Equipment Selection

In this chapter:

- *Is Rebreather Diving Right For You?*
- *How to Select the Right Rebreather*
- *Auxiliary Equipment*
- *Commonly Available Tanks*
- *Consumables*
- *Basic Rebreather Care*

Is Rebreather Diving Right For You?

Now that you know a little bit about rebreathers, it is time to ask yourself if rebreather diving is right for you. This is the point where I get to regale you with cool stories about the possibilities offered by rebreathers but add a pinch of reality. I have buried a lot of friends because of accidents they had on rebreathers. Those losses have left me saddened and forever changed. I'm going to share some sad stories, because I believe we can't gloss over the bad stuff. This is a huge decision for you, your family and your heirs. If you choose to take on more risk in your diving, you better have all the facts.

Your informed decision is going to come with some sober details from your new big sister, Jill.

The Good, Bad and the Ugly

When I met Dr. Bill Stone, I could feel an immediate and magnetizing draw towards the exploration potential of rebreathers. They were incredibly alluring to me. I knew the technology was capable of significantly extending the limits of my dives and it presented an opportunity for new interactions with wildlife. I envisioned myself swimming to the end of every cave line in Florida and pushing out the current exploration. I pictured myself hanging out with gentle humpback whales as a fellow traveler in the ocean, no bubbles to disturb our connection. I loved technology and problem solving and this new apparatus would allow me to tinker and create tools for fascinating projects all over the world. I'll be honest. I

CHAPTER 3 - EQUIPMENT SELECTION

did not consider the additional risk at first. I was too mesmerized. My former husband Paul and I were very experienced divers. We were also cautious. Surely, mastering this new equipment could be done in a safe, methodical and fun way. Our biggest concern was making the $35,000 investment to land two of these things in our west Florida dive shop. Paul and I actually got our first rebreathers as loaners from Jim King who was sponsoring the Wakulla2 United States Deep Caving Team Project. Jim purchased six of the new Cis-Lunar MK-5P rebreathers and had them shipped to our dive shop. The units were provided for potential project exploration divers for training purposes. Paul and I would provide a home for the units, learn them, teach them to other participants and act as "training central" for the upcoming project. Over the course of two years, explorers from all over the world stayed with us and

Wary of new technology, we carried onboard double 104 cft bailout and double 80 cft aluminum sidemount tanks on deep dives. Photo: U.S. Deep Caving Team.

dived the loaner units. These guests lived with us in a ramshackle single-wide trailer, made technological masterpieces in our workshop, fell through our withering bathroom floor and totally enriched our lives. But early on, Paul and I decided that we were not interested in sharing such an important piece of equipment, even if they were provided free for our use. We were immediately hooked and decided to make in investment in purchasing our own personal rebreathers.

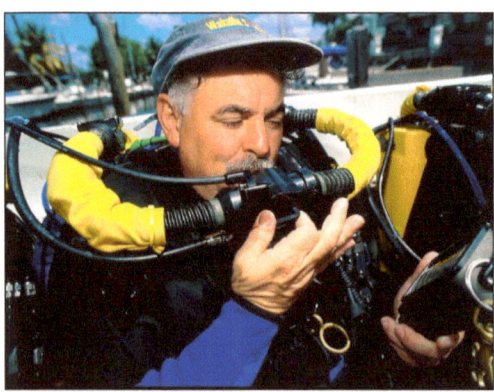

Paul Heinerth finishing up his pre-dive check on his Cis-Lunar CCR. Note the large console we called "the brick." It contained a primary computer on one side and an independent bulkheaded analog redundant oxygen display (ROD) on the back of the heavy unit. Photo: Richard Nordstrom.

With no formal instruction at first, Paul and I tested, learned and got feedback from Bill in Maryland - and experimented some more. They say that it is the stuff

CHAPTER 3 - EQUIPMENT SELECTION

you don't know that could kill you. This is more true than you can imagine in CCR diving. When I look back on those first years of learning, I can say I am incredibly fortunate to be writing this book today.

I recall one experiment held in the classroom of our SCUBA shop. Andrew Poole and John Vanderleest were visiting from Australia. They were staying with us and training on the units, preparing for the Wakulla2 Project. Like us, John and Andrew had experience with Dräger rebreathers. They had been using them in Australia's cave systems. At the time, the Cave Diving Association of Australia (CDAA) was tightly regulating conduct in the country's known cave systems. Nitrox use was strictly banned. Only air diving was permitted. There was no specific policy on rebreathers at the time. One day, the leadership in the organization decided that by virtue of the fact that the rebreather was mixing gas, then they were using nitrox to dive. John and Andrew were banned from entering the CDAA-permitted caves for a period of one year. After emailing me about participating in Wakulla2, I urged them to come stay with us and learn the MK-5P rebreathers.

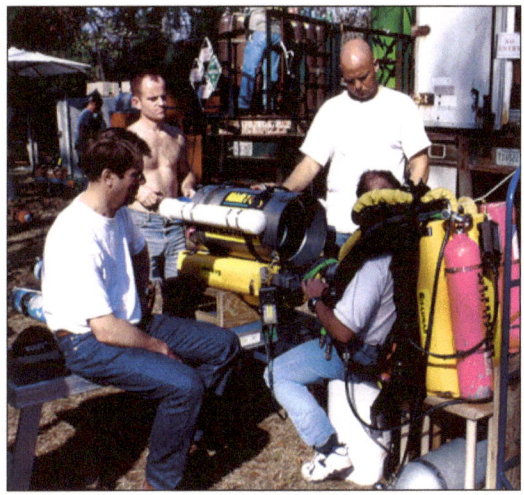

Wakulla2 exploration divers (left to right), John Vanderleest (AUS), Jason Malinson (UK), Matt Matthes (Mexico), Andrew Poole (AUS), working on a scooter-mounted bailout rebreather at Wakulla Springs in 1988 (above). Experimenting with the dual-scooter-mounted rebreather at Wakulla (below). Photos: U.S. Deep Caving Team.

Their background would be an asset to the project. This became quite a joke in our cave diving circles. Their punishment for breaking the rules in Australia would be participation in the most technologically advanced cave diving exploration (up to that point) in history.

The 1990s were very different times than today. When I lived in the Cayman Islands in the early 90's, nitrox use was banned there too. In fact, an entire issue of Skin Diver magazine had been dedicated to explaining why various top professionals would never use the dangerous voodoo gas – nitrox. In the earliest days of the sport CCR diving, the same warnings of impending disaster were being circulated about rebreathers. Some of that doom and gloom has persisted within the diving community. Sage divers with no rebreather experience are likely

CHAPTER 3 - EQUIPMENT SELECTION

to tell you that you are going to kill yourself. But, I want you to have the facts, consider them carefully, talk them over with your family and then make an informed decision.

Now, back to the experiment. John, Andrew and I were trying to get our heads around every single thing that could go wrong on a rebreather, then develop good protocols for dealing with them. We spent a lot of time talking about hypoxia. John had experienced an incident with hypoxia when the orifice on his Drager rebreather became restricted and caused him to blackout in a cave. He survived the event due to the quick action of his buddy Andrew. The incident obviously lingered in their minds. What problems in the Cis-Lunar could induce hypoxia? Could a diver detect the problem in time to prevent a blackout?

We decided to conduct a very risky but voluntary experiment. Andrew would set up a video camera to film the activity. We had a DAN oxygen kit ready to use. One of us would sit on the floor in the MK-5P rebreather. The subject would hold a notepad and paper but could not see any of the the displays that would normally alert them to trouble. We would turn off the oxygen supply but give the diver sufficient volume to breathe. The "victim" knew that his oxygen was slowly being depleted in the experiment. He was instructed to write down any symptoms that he felt that might indicate a problem and he was told to "bailout" by turning a switch on the front of the mouthpiece block when he thought he was experiencing hypoxia. The question we were trying to answer was "can you feel hypoxia coming on and prevent your own blackout?"

You've likely already figured out that telling someone to write down symptoms is pretty contradictory to the action of bailout. If the victim feels compelled to write

The final design of the dual redundant Cis-Lunar MK-5P rebreather.

CHAPTER 3 - EQUIPMENT SELECTION

something down, he should be trying to save his life by turning the bailout valve to get safe gas. Right? Well it appeared it was not that easy to save your own life.

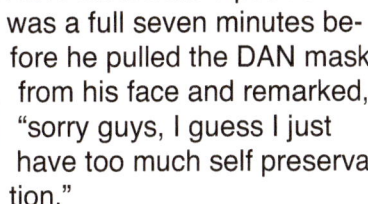

I watched Andrew carefully and monitored his displays while John manned the camera. I turned off the oxygen supply and told Andrew we were ready to begin. He nodded and smiled. It took ages for the PO_2 to drop. From time to time, the reduced volume would trigger the automatic diluent valve and fresh diluent would flow into the loop. I watched the numbers tick down. First I saw 0.18. He looked normal other than the fact that the mouthpiece block was feeling heavy out of the water. Then 0.16. This should be the threshold of consciousness. Apparently not. He lifted his hand slightly, looked puzzled and set it down again. He continued to breathe as the minutes rolled on. It read 0.14 and he was still breathing. His color was slightly ashen and he picked up the pencil. "Tingling," he wrote in a messy script on the yellow ruled paper pad. I looked up at John to let him know we were close. I checked the display to see it read 0.12 and then it dropped another tenth. He placed the pencil awkwardly and wrote, "T..W," then his hand spasmed and he slumped forward in a weird chaos of twitching. I yanked the block from his mouth and threw the oxygen mask on his face, hitting the demand valve and sending him a boost of rich, fresh gas. He convulsed. It was terrifying. My new friend was wearing a mask of death. For moments I envisioned that this stupid experiment had just killed Andrew, while on the other side of the SCUBA West wall, people were shopping for new colorful dive skins. It was horrible. Slowly he came around and shuffled himself into a more comfortable position. It

was a full seven minutes before he pulled the DAN mask from his face and remarked, "sorry guys, I guess I just have too much self preservation."

I replied, "whatever," and pushed the mask back over his face.

A few more minutes passed and he got a confused look on his face, "guys, what just happened? I thought I bailed out?"

CHAPTER 3 - EQUIPMENT SELECTION

There are a lot of morals to this story. The most important is don't try this at home. People who develop hypoxia might not come around when you stick an oxygen mask on their face. But on that day, another lesson was clear to me. Andrew was expecting hypoxia. He even had the cognitive ability to note symptoms. But it was already too late. He might have appeared conscious, but he was long gone. He was disabled for many minutes prior to blackout. Andrew actually recalled recognizing the symptoms and thinking, "gee, it's time to bailout." He actually thought he had bailed. But his mind never managed to get his hands to initiate the task of bailout. If he had been underwater, he would have died.

The author prepares to descend with dive partner Dr. Tom Iliffe on the deepest dive ever conducted in Bermuda during the NOAA Deep Water Caves project in Bermuda in 2011. Photo: Nic Alvarado.

I'd never do an experiment like that again, but I am glad to have had the experience. It scared the life out of me and gave me a new respect for monitoring my rebreather's displays. These boxes could indeed be silent killers that could take the lives of friends. The classroom experience left us all deeply introspective. Andrew was shocked when he saw himself on tape. Few people get a chance to gaze at their own impending death.

And so I began my rebreather career, always maintaining a healthy fear of the box that could kill me. I no longer had stars in my eyes. This technology would undoubtedly lead to incredible experiences but only if I gave it due respect. I decided I was the type of diver that could give it respect.

Mandy Gurr with daughters Amberlee and Leyla making rebreather diving a family project. Photo: Courtesy of Kevin Gurr.

So, how do you know if you are the right kind of diver and if you are capable of offering the proper amount of respect to the technology?

Each training agency and manufacturer has crafted a list of prerequisites for safe rebreather diving. They'll suggest that you have a nitrox qualification and

CHAPTER 3 - EQUIPMENT SELECTION

can display good buoyancy control. They'll suggest that you review basic physics and be comfortable calculating partial pressures. They'll require a number of prerequisite dives. However, if you want to be a rebreather diver, you just can't tick off "been there, done that" boxes. You have to look much harder at your attitude and behavior. You must be meticulous. You must be prepared to follow strict protocols and continue using those safety protocols every time you dive, in every condition, every day, without question. You have to promise that you will never dive your rebreather if something is not functioning perfectly. You have to promise your wife, husband, kids, friends and any higher power you believe in. If you can do this, then you might be a good candidate for diving a rebreather. If you can't be diligent and disciplined, you stand a reasonable chance of dying on one.

How to Select the Right Rebreather

When you ask someone, "what is the best rebreather?" they will undoubtedly describe the one they own. Divers develop a strong attachment to their shiny, new and significant investment, and are unlikely to point out any failings in their gear. They may even be right. Their rebreather may indeed be the best rebreather… for them.

The bottom line is that the last decade of statistics shows us that the actual equipment brand does not seem to be a factor in accidents. Best available stats show us that there are no more accidents on one particular brand than another based on ratios to the number of units sold. The same appears to be true when we look at manual versus electronic units. That means, the most important decision you can make is not "which" rebreather you buy, but rather "how" you will use it.

There are a lot of things that should affect your purchase decision, even before you start to look at individual features.

Third-Party Testing. Remember the little red and back circles in the Consumer Reports Magazine that highlight faults in various household products? We don't have that for rebreathers. Much of the manufacturing process is proprietary and secretive. Third-party testing and validation ensures that quality control standards have been applied in the manufacturing process and that industry-agreed-upon safety standards are met consistently. You should insist on seeing CE or equivalent test data for work of breathing, oxygen tracking, canister duration and other factors. See Chapter 10 for details.

CHAPTER 3 - EQUIPMENT SELECTION

Portability. If you are planning to travel with your unit, will it fit within baggage weight standards for your carrier? Is it modular? Is it common enough that you can find tanks and other parts at a vacation destination? Is there international support for your rebreather?

Instruction. You'll want to look for an instructor even before you buy a unit. The instructor base for many rebreathers is still rather small. It is important to learn from a trusted, experienced and current instructor. The manufacturer may offer a "qualified" list of instructors that they recommend and that have a proven background and currency (frequent diving as opposed to dollars!) enough to gain their recommendation. You may need to budget for travel to an instructor's home base in order to get introductory or advanced training.

Support. Having local support and role modeling from dive shops or experienced locals on a particular unit is worth its weight in gold. If everyone in your dive club owns one type of rebreather, then you should give it a hard look. Having role models, spare parts and dive boat support is very helpful.

Service. Do a little research on the manufacturer and their support centers. Can you get local service or do you have to ship internationally for repair? Does the manufacturer or support center offer good customer service, timely repairs and reasonable warranty coverage? Will shipping delays cause you to miss a season of planned dives?

The author building rebreathers for the Hollywood feature movie, "The Cave." Two crew members were trained to build rebreathers from the ground up in order to handle all service on site during filming in Romania and Mexico. Photo: Leon Scamahorn, Innerspace Research.

Self Service. If you are a traveler, are there easy things that you can repair or replace or are the parts and tools all proprietary and complex? Is this important to you? Some rebreathers can be serviced by local CCR dive shops but others must be sent to the manufacturer. Some people want to be able to quickly replace sub-components on an expedition and others may never want to self-service anything.

History. How long has the manufacturer been in business? Is the business reliant on a single person as the company brain trust or is there a larger entity behind the brand? When software engineer Will Smithers tragically died in a helicopter crash, his computer brilliance was also lost. His intellectual property powered a number of early CCRs. You'll want to learn a little bit about the manufac-

CHAPTER 3 - EQUIPMENT SELECTION

turer of your rebreather and whether you feel like the company is stable and robust and likely to be around for the life of your purchase.

Budget. Your budget will affect your choice of rebreather. You may want to own the top of the line technical rebreather that makes coffee at the dive site, but have to look realistically at a lower price point. You need to budget for the unit itself, proper instruction, consumables and regular maintenance. Don't blow the budget on the unit itself and then cut corners on replacing sensors. If you don't have the finances to support meticulous maintenance, then don't buy the unit in the first place.

> **SAFETY CHECK**
>
> Use only the absorbent brand that is recommended by the manufacturer. Different brands and grades may change canister durations by up to 50%.

Availability of Consumables. You will regularly use batteries in your rebreather. Some are rechargeable and some require unique replaceable batteries. You will also regularly use carbon dioxide absorbent in the form of either granular sorb or pre-packed canisters. Are these easily available in your region and the places you want to travel? You will also need to periodically change out oxygen sensors. Can you get fresh sensors locally? Sensors have a shelf life whether they are in service or not, and if you can get them locally it might save you from throwing away unused sensors that you purchased, but never installed because they are past their expiration date.

Can you consider a used rebreather? If you find a great deal, you should budget on sending the unit back to the manufacturer to restore it to factory specifications and performance. You'll also want to start with fresh sensors. This can easily add up to several hundred if not over a thousand dollars. Be sure the deal is still worth it.

It's a tough decision and nearly everyone you ask for an opinion will have a strong one. They just bought the equivalent of a Harley-Davidson Sportster motorcycle, and will likely ooze with confidence about their buying decision. You have to be pragmatic and analytical. A try-dive experience will only help you fall in love with a harness. A comparison of cosmetics will only help you decide whether you look cool. Take time to compare features and research the unit on your own before jumping into one of the biggest purchase decisions of your underwater life.

CHAPTER 3 - EQUIPMENT SELECTION

Comparing Rebreather Features

There are numerous features available on various rebreather brands. Use the following checklists to help you compare rebreather features and function.

Comparing Warning Systems

Does the unit have any built-in alarm systems? Are they visual, audible or tactile alarms and where will the diver see the actual warning? Does the particular model offer any of the following warnings?

- ✓ Audible alarms
- ✓ Lights on HUD for alarms
- ✓ Primary handset flashes
- ✓ Handset color change for alarms
- ✓ Tactile, vibrating alarm
- ✓ Pre-dive check incomplete
- ✓ Pre-breathe incomplete
- ✓ Pre-dive aborted
- ✓ Pre-dive failed
- ✓ PO_2 okay
- ✓ High PO_2
- ✓ Low PO_2
- ✓ Solenoid failure
- ✓ Improper setpoint
- ✓ Calibration needed
- ✓ Oxygen cells okay
- ✓ Oxygen cell(s) disabled
- ✓ Cell error
- ✓ Scrubber okay
- ✓ Thermal profile okay
- ✓ Scrubber bad
- ✓ Thermal profiler turned off
- ✓ Absorbent time expended
- ✓ High carbon dioxide
- ✓ Carbon dioxide okay
- ✓ CO_2 monitor turned off
- ✓ Oxygen tank off
- ✓ Diluent tank off
- ✓ Low oxygen pressure
- ✓ Low diluent pressure
- ✓ Oxygen empty
- ✓ Diluent empty
- ✓ Pressure reading out of range
- ✓ Rapid diluent use (for leaks)
- ✓ Rapid Oxygen Use (for leaks)
- ✓ Decompression required
- ✓ At stop depth
- ✓ Below stop depth
- ✓ Depth of diluent exceeded
- ✓ Too deep for unit
- ✓ Fast ascent
- ✓ Battery okay
- ✓ Battery low
- ✓ Battery failed
- ✓ Abort
- ✓ Diluent flush now
- ✓ Service needed
- ✓ Other alarm features

CHAPTER 3 - EQUIPMENT SELECTION

Comparing Automation/Electronics

Compare the following features between models you are considering. Are these features important to you?

- ✓ Auto-on when breathing detected
- ✓ Auto-on at depth
- ✓ Auto-on when wet
- ✓ Primary handset
- ✓ Secondary handset
- ✓ HUD
- ✓ Buddy HUD
- ✓ Automatic setpoint mode (increments with depth)
- ✓ Auto setpoint switch at specific depths
- ✓ Dual set point mode
- ✓ Automated oxygen sensor disable
- ✓ Over-ride to auto sensor disable
- ✓ OLED display
- ✓ Other highly readable display
- ✓ Readable type size on display
- ✓ Onboard decompression
- ✓ User selectable deco algorithm
- ✓ Independent secondary PO_2 meter
- ✓ Deco option on secondary
- ✓ On screen pre-dive checklist
- ✓ Absorbent timer
- ✓ PC download
- ✓ Simulator
- ✓ Dive planner
- ✓ Dive logger
- ✓ Rechargeable batteries
- ✓ Life support (PO_2 control) separate from display units
- ✓ Deco status on HUD
- ✓ PO_2 status on HUD
- ✓ Calibration checks at depth
- ✓ Linearity checks at depth
- ✓ Other automation features

CHAPTER 3 - EQUIPMENT SELECTION

Comparing Mechanical Features

The price of a rebreather may come down to the features and luxury items that it offers. Compare the list of features below to decide whether these items are critical, wanted or simply an added bonus.

- ✓ Depth rating
- ✓ Mixed gas approved
- ✓ CE tested rebreather
- ✓ CE rated Bailout Valve (BOV)
- ✓ CE approved regulators
- ✓ BOV isolator
- ✓ CO_2 sensor (breakthrough sensor)
- ✓ Thermal array for canister (duration estimator)
- ✓ Metabolic oxygen counter (duration estimator)
- ✓ Absorbent timer (duration estimator)
- ✓ Manual addition valves
- ✓ Manual bypass gas blocks
- ✓ Backmounted counterlung
- ✓ Front mounted counterlung
- ✓ Over-the-shoulder counterlung
- ✓ Automatic Diluent Valve (ADV)
- ✓ ADV isolator
- ✓ Oxygen solenoid isolator
- ✓ High pressure (10 bar +) O_2 solenoid
- ✓ Offboard gas connection kit (routes through automation)
- ✓ Hydrostatically balanced OPV
- ✓ Visual alarms
- ✓ Audible alarm
- ✓ Tactile alarm
- ✓ User replaceable display units
- ✓ User packed scrubber
- ✓ Pre-packed scrubber
- ✓ Other features

Comparing Physical Characteristics

These features describe the basic build of the unit itself. They include ergonomics and features that you may be able specify at time of purchase.

- ✓ Weight
- ✓ Buoyancy
- ✓ Packing case (included?)
- ✓ Weight in packing case
- ✓ Onboard tanks (steel or aluminum)
- ✓ Harness (included?)
- ✓ Harness type
- ✓ Wing (included?)
- ✓ Wing style (backmount, BCD, etc.)
- ✓ Overall fit and comfort
- ✓ Modular for packing
- ✓ Hard case or cover
- ✓ Stand for easy dressing
- ✓ Other features

CHAPTER 3 - EQUIPMENT SELECTION

Comparing Support

Never underestimate the importance of how and where you can get support and parts for your rebreather. This may include a little research online to uncover whether other owners feel like they are getting good customer support and warranty on their purchase.

- ✓ Spare parts kit
- ✓ Good user manual
- ✓ Good user service manual
- ✓ Good educational materials
- ✓ Pre-dive checklist
- ✓ Build checklist
- ✓ Post-dive checklist
- ✓ Service checklist
- ✓ Online support
- ✓ Local service
- ✓ Company viability
- ✓ Instructor choices
- ✓ Travel weight okay
- ✓ Travel support
- ✓ Scrubber material availability

Auxiliary Equipment

Besides the rebreather itself, there are a number of other pieces of dive gear you will need to gather together or purchase before beginning a class. Following are some recommendations and tips for accessory equipment.

Open Circuit Bailout Equipment

Don't sell all your open circuit gear yet! You are going to need at least one open circuit bailout tank to carry with your rebreather. Some recreational divers who are diving at depths shallower than 60 feet/20 meters, may have enough gas in their onboard tank to get them safely to the surface, but you will find that most prudent divers carry bailout for themselves and/or to assist a buddy diver that needs gas. At a minimum, you must have an alternate air source (second stage) to use or offer to another diver. That may or may not be included with your rebreather.

The volume of your bailout tank(s) must be adequate to get you safely to the surface on open circuit with a sufficient margin for error. If you plan on diving in overhead environments or

Don't forget to budget for auxiliary equipment such as a bailout tank and regulator.

43

CHAPTER 3 - EQUIPMENT SELECTION

plan on decompression diving in the future, then that open circuit gas supply must be adequate to get you through an exit and decompression. Technical CCR divers may carry or stage multiple tanks for emergencies. Their deepest offboard tank must be filled with a safe gas that can be breathed at maximum depth. They may carry or stage other mixes to optimize an open circuit decompression obligation in the event they ever suffer a catastrophic loop failure that forces them to abort on open circuit.

Before you sell any of your old open circuit gear, consider the type of diving you might like to do several years down the road. Keep enough tanks and regulators to cover your bailout needs. Ensure that those regulators are of sufficient quality for deep diving, if that is in your plans. One of the emergencies that may force you onto open circuit is a carbon dioxide emergency. In the event of high CO_2, you may already be breathing at an accelerated rate. You need to be able to bailout to a safe breathing mix on a well-tuned, CE-rated regulator that is designed for deep diving. If the regulator on your bailout tank is of poor quality or poorly tuned, then you may not be able to overcome the rise in carbon dioxide in your body. This could lead to blackout and drowning. Prevent this by purchasing and maintaining good equipment.

Gil Nolan preps a stage/bailout bottle with standard rigging (above). A butt plate is used to move the bottom tank attachment lower and further around the hips to better streamline the tank (below).

Your bailout tank should include a first stage, second stage, and submersible pressure gauge (SPG). It may also carry a dry suit inflator hose. In some cases a special connector whip links the offboard bottle to the rebreather. These should be mounted in a streamlined fashion with hoses carefully stowed under bungees or tire inner tubes. The second stage needs to be easy to access,

CHAPTER 3 - EQUIPMENT SELECTION

but well-stowed so it does not drag or come loose. A stage tank kit can be used for securing the tank or you can make one yourself. The tank should be trimmed close to your body in a streamlined fashion and should not hang perpendicular to your body where it can drag on a reef or get damaged. Bailout tanks should be properly analyzed and marked. You should also check that your buddy's bailout is properly analyzed and marked in case you need to use it.

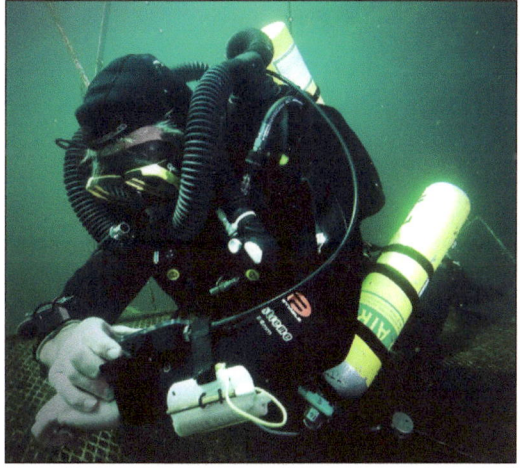

You have a choice between aluminum and steel tanks for bailout. Some divers prefer aluminum because they are light, almost neutral in the water and easy to remove and replace due to their buoyancy characteristics. A steel tank of a comparable size often carries significantly more volume. If you are diving in cold water, if you use a dry suit or need some lead weight for buoyancy control, steel bailout tank(s) may be your best choice. I prefer carrying my weight as usable gas rather than as lead weight. I dive in caves most of the time and may need a significant volume to get out of the cave in an emergency. I have chosen to use two 45cft steel tanks when I dive my rebreather with my drysuit. That leaves me perfectly neutral without the need for lead. Additional stage bottles

An aluminum bailout bottle clipped on a chest D-ring will fall into a position that is perpendicular to the diver's body. The regulator may drag on the bottom, damaging equipment or environment (above). By using a bungee system, the tank is pulled up and back and hangs in a streamlined fashion parallel to the diver's body (below).

for deep diving are aluminum since I don't need any more negative buoyancy. In a tropical environment in a light wetsuit, I switch over to aluminum tanks. This is convenient for travel since aluminum tanks are often the only thing available. Many recreational rebreather divers choose to use 30 (4.3l) or 40 (5.7l) cft tanks.

If you use a single bailout tank, then you may need to counterweight it slightly by putting a very small weight on your harness or weight belt on the opposite side. If you fail to counterweight you may feel off balance in the water. Before buying any new tanks, borrow or rent different sizes and styles and try them out until you get your buoyancy trimmed perfectly.

CHAPTER 3 - EQUIPMENT SELECTION

Common Tanks

The following charts list most commonly available cylinders and their weight and size specifications. Standard onboard and commonly selected bailout tanks are shown in darker colored columns. If your tank is not listed here, ask the manufacturer for similar specifications to help you predict buoyancy changes if you switch to another brand or model of tank. Imperial measurements are shown in the top half of the chart in green. Metric measurements are shown in the bottom half of the chart in orange. All buoyancy has been calculated in sea water with a valve present but not a regulator.

WORTHINGTON	LP27	LP77	LP85	LP95	LP108
Service Pressure (psi)	2400+	2400+	2400+	2400+	2400+
True Capacity @ 2640 psi	27 cft	77 cft	82.9 cft	93.3 cft	108.6
Buoyancy Full (lbs)*	-3	-6.8	-7.1	-10.1	-10.7
Buoyancy Empty (lbs)*	-1	-1	-0.7	-3	-2.6
Weight Empty	11.9	32.5	36.8	41.9	45.9
Outer Diameter (inches)	14.5	23.2	24.7	23.7	26.5
Length (inches)	14.5	23.2	24.7	23.7	26.5
Service pressure (bar)	166+	166+	166+	166+	166+
Water Capacity (liters)	4.3	12.0	12.8	14.8	16.8
Buoyancy Full (kg)*	-1.4	-2.7	-2.7	-3.6	-3.6
Buoyancy Empty (kg)*	-0.5	-1.0	-1.0	-1.5	-0.5
Weight Empty	5.4	15.6	16.7	18.5	20.2
Outer Diameter (mm)	140	184	184	203	203
Length (cm)	36.8	65.3	65.3	60.1	67.9
Common Use	bailout				

CHAPTER 3 - EQUIPMENT SELECTION

XS SCUBA	X7-65	X7-80	X7-100	X7-120	X8-119
Service Pressure (psi)	3442	3442	3442	3442	3442
True Capacity @ 2640 psi	66.4 cft	81 cft	99.5 cft	120.6 cft	123 cft
Buoyancy Full (lbs)*	-8.7	-9	-10	-11	-10.9
Buoyancy Empty (lbs)*	-3.9	-3	-2.5	-2	-2
Weight Empty	25.1	29.9	33.1	39.7	42.5
Outer Diameter (inches)	7.25	7.25	7.25	7.25	8
Length (inches)	16.7	19.8	22.7	27.7	23.9
Service pressure (bar)	230	230	230	230	230
Internal Volume (liters)	8.2	10.1	12.2	15.3	14.8
Buoyancy Full (kg)*	-3.9	-4.1	-4.5	-4.9	-4.9
Buoyancy Empty (kg)*	-1.8	-1.4	-1.1	-0.9	-0.9
Weight Empty	11.4	13.1	15	17.2	19
Outer Diameter (mm)	184	184	184	184	203
Length (cm)	42.2	50	60.1	71.1	60.1

CHAPTER 3 - EQUIPMENT SELECTION

BLUE STEEL FABER	L27DV	L50DVB	L85DVB	M71DVB	M100DVB
Service Pressure (psi)	2640	2640	2640	3300	3498
True Capacity	27 cft	50 cft	85 cft	71 cft	100 cft
Buoyancy Full (lbs)*	-3	-2.4	-3.8	-4.4	-14.1
Buoyancy Empty (lbs)*	-1.1	-1.2	-2.3	-0.9	-6.7
Weight Empty	11.7	18.9	31.2	28.7	38.7
Outer Diameter (inches)	5.5	5.5	7	6.8	7.25
Length (inches)	14.4	25.2	26	20.5	24.2
Service pressure (bar)	180	180	180	225	240
Internal Volume (liters)	4	8	13	9	12
Common Use	SCR	bailout			

Rebreathers are inherently bulky and often boxy pieces of equipment. Divers are frequently so focused on the rebreather itself, that they fail to take the time to find a method that streamlines their bailout gas. As you've likely realized, from the number of pages I am dedicating to bailout tanks, this is one of my pet peeves. In the photos above, you can see how the diver's tank (left) hangs perpendicular to the bottom in poor trim. The diver on the right has a much more desirable trim. The tank valve lays under his armpit and the cylinder is inline with his body, offering streamlined trim. Attention to detail protects equipment and the environment, lowers workload and lessens the likelihood of entanglement.

CHAPTER 3 - EQUIPMENT SELECTION

One of the skills you will learn in class is the removal and replacement of your bailout tank while stationary and swimming. Aluminum tanks are more neutrally buoyant than steel tanks and are simple to put on in the water. Make sure the clips that you use on your bailout tank are easy to open and close with whatever hand protection you will wear. Dry gloves or heavy neoprene mitts demand that you use reasonably large clips. Make sure you can reach and easily operate all clips.

LUXFER ALUMINUM	AL06	AL13	AL14	AL19	AL30
Service Pressure (psi)	3000	3000	2015	3000	3000
True Capacity	6.1 cft	13.2 cft	13.7 cft	19.9 cft	30 cft
Buoyancy Full (lbs)*	-1.4	-1.7	+0.8	-1.4	-1.5
Buoyancy 500 psi (lbs)*	-1	-0.9	+1.6	-0.1	-0.4
Weight Empty	2.5	6	4.9	8.2	12.3
Outer Diameter (inches)	3.2	4.4	4.4	4.4	5.25
Length (inches)	11.1	13.1	16.2	18.6	19.3
Service pressure (bar)	207	207	139	207	207
Internal Volume (liters)	0.9	1.9	2.8	2.9	4.3
Buoyancy Full (kg)*	-0.6	-0.8	+.4	-0.6	-0.5
Buoyancy Empty (kg)*	-0.4	-0.3	+.8	+.05	+.5
Weight Empty	1.2	2.7	2.2	3.7	5.3
Outer Diameter (mm)	81	111	111	111	124
Length (cm)	28.3	33.3	41.1	47.1	55.5
Common Use	onboard	onboard		onboard	bailout

CHAPTER 3 - EQUIPMENT SELECTION

LUXFER ALUMINUM	AL40	AL50	AL63	AL80	ALN80
Service Pressure (psi)	3000	3000	3000	3000	3300
True Capacity	40 cft	48.4 cft	63 cft	77.4 cft	77.4 cft
Buoyancy Full (lbs)*	-2.3	-2.2	-1.4	-5.7	-4.3
Buoyancy 500 psi (lbs)*	+1.4	+0.8	+1.7	+3.4	-0.9
Weight Empty	21.2	26.7	31.4	35.4	41
Outer Diameter (inches)	6.9	7.25	7.25	7.25	8
Length (inches)	19	21.9	26.1	25.8	26.2
Service pressure (bar)	207	207	207	207	228
Internal Volume (liters)	5.7	6.9	9	11.1	10.3
Buoyancy Full (kg)*	-0.3	-1	-0.6	-2.6	-2
Buoyancy Empty (kg)*	+1	+.6	+1.2	+.2	+.02
Weight Empty	6.9	9.6	12.1	14.2	16.1
Outer Diameter (mm)	133	175	184	184	184
Length (cm)	62.9	48.3	55.5	66.2	65.6
Common Use	bailout	bailout	bailout	bailout	

Exposure Suits

There are unique challenges to learning good buoyancy control when you start diving a rebreather. You should use the appropriate thermal protection for the diving environment where you will be taking your class. The rebreather itself generates an exothermic reaction in the scrubber canister. You will be a little warmer than you are accustomed to, since the scrubber will be pumping out heat and moisture.

A rebreather class is not the place to try out a new suit. If you are diving dry, ensure that you are completely competent in dry suit diving. There are some advantages to diving dry. You may find it easier to get well trimmed in the water since you will be able to shift the air around in the suit to the region where lift is required. A wet suit won't work in the same way, but you also don't have to deal with venting your suit on ascent. If you dive wet, you may want to bring your full

CHAPTER 3 - EQUIPMENT SELECTION

collection of boots and bottoms. Counterlungs in rebreathers tend to perch a bubble of gas high on the shoulders, causing feet to drag along the bottom. In order to lift your feet, you can try thicker neoprene boots, added neoprene socks and neutral or positively buoyant fins. It is going to take some patience and craft to get you swimming in a comfortably trimmed position.

If you are diving dry, your LP supply hose may be put on your bailout tank or an inflation bottle, if that is what you are accustomed to. You will not likely attach it to the onboard diluent supply, since that tank is minimal in volume. If using an inflation bottle, it is generally mounted with the tank valve down on the left side of the body in such a way that it cannot be mistaken for the diluent tank valve.

Writing Tools

You are going to have a lot of time underwater while you accumulate hours during class. It is extremely helpful to carry a pencil and wrist slate or notebook. You'll have lots of complex questions for your instructor underwater and you'll want to take notes regarding drills and consumables. You can effectively project your voice and talk through a rebreather, but you also want to keep track of information and questions for after the dive.

There are times when you will use a Delayed Surface Marker Buoy (DSMB) for ascent. The lift bag and reel can be stowed on the hip or butt D-ring. When you are ready to deploy it, hold the spool or reel well away from your body to prevent entanglement. When it hits the surface, you can easily hang suspended on your safety or decompression stop.

CHAPTER 3 - EQUIPMENT SELECTION

DSMB

Boats find it difficult to keep track of rebreather divers because there is no telltale bubble trail streaming off your rebreather. A Delayed Surface Marker Buoy (DSMB) is a brightly colored inflatable float used by divers to mark their position. DSMBs are used for ascent and drift dives and you will learn how to deploy these proficiently during a rebreather class. If you are competent in their use in open circuit, it is going to feel quite different on a CCR. There are many ways to deploy various styles of DSMBs and your instructor will coach you on the best selection and technique. Bear in mind that you need to be very careful not to become entangled in an ascending lift bag. If you become entangled in a lift bag that carries you upward you may lose buoyancy control, miss a decompression stop or even reach a dangerously low PO_2 very rapidly. Practice in shallow water and stay proficient in their use.

Spools are often preferred over using reels with DSMBs. They are generally lower profile and can be deployed with only a rare risk of jamming.

Backup Instrumentation

Depending on the type of rebreather you choose, you may be fortunate enough to have two onboard independent computers that monitor rebreather status as well as depth, time and other functions. If you have chosen a manual rebreather, there may be no onboard decompression information at all. Each diver should have a primary device as well as a backup. Several manufacturers make independent computers that can be used with rebreathers. Some are very simple devices that can be used to track setpoint and others may be plugged into the rebreather to monitor actual setpoint or an independent fourth oxygen sensor. Some divers choose to use traditional depth gauges and bottom timers with tables, although that does not allow a CCR diver to take full advantage of the extended range that the rebreather offers. SCR divers may choose to use an open circuit nitrox computer set conservatively for an equivalent percentage of nitrox.

CHAPTER 3 - EQUIPMENT SELECTION

Cutting Tools

You should carry a cutting tool (preferably two). Many divers choose to carry some sort of knife and a pair of underwater shears. These devices may be used to cut away entanglements or free a fouled reel line.

Mask, Fins and Snorkels

Your mask needs to fit exceptionally well and you may need to read small characters on the onboard controller, requiring a mask with vision correction. Give yourself a reading test with your mask before your class in case there are any issues. Fit is critical because you have to become a "molecule miser." Every bubble of gas is critical and you can't afford to be continually leaking through a poorly sealed mask. Furthermore, if you have to clear your mask often, you will find that each time you do so, you will lose buoyancy. When you vent bubbles from your mask, you lose overall buoyant volume. That has to be replaced with gas that reaches your setpoint. Sometimes you get a vicious cycle of overuse. First, you clear your mask and lose buoyancy. Then you add diluent gas that recovers buoyancy, but now you have arrived at a lowered PO_2. With the PO_2 drop, the oxygen solenoid fires. When that fires a few times to raise the PO_2, then buoyant volume expands. If the gas volume is too great, then you will vent through your OPV. As a result, one small mask clearing exercise has become a frustrating buoyancy seesaw that uses a lot of gas. During your training, you will learn strategies for mitigating these swings in buoyancy, but for now, the best solutions are to make sure your mask has a great seal and to confirm that you are not somebody who regularly vents gas through their nose.

As mentioned earlier, sometimes negatively buoyant fins become a challenge when the CCR tends to pitch your body to a shoulder-high position. Be open-minded about getting another pair of fins that are lighter in the water. The mass of the rebreather may be a little greater than your standard gear. It may create a greater drag in the water. Streamlining and a good pair of power fins will help you move this mass through the water.

Snorkels are not very convenient to wear with a CCR, as the loop gets in the way. Many rebreather divers carry a collapsible pocket snorkel so that they are not left without one on the surface. It's safer to be off the loop and on a snorkel while you swim on the surface to the boat rather than staying on the rebreather.

CHAPTER 3 - EQUIPMENT SELECTION

Several unfortunate accidents have occurred when divers disabled their system and swam to the boat without adding sufficient oxygen.

Full Face Masks

Full face masks (FFM) are extremely specialized pieces of equipment and are not suitable to use when you are learning to dive a rebreather. Filmmakers sometimes use them to capture clean vocals when filming underwater and commercial divers use them for routine communications. FFMs have the benefit of lessening the likelihood of drowning if you become unconscious, but there are other significant risks to consider before using one. I have used FFMs on film shoots countless times. This is purely for the purpose of recording audio or being able to communicate with my crew. Some people argue that a FFM can save a diver that passed out due to hypoxia. That may be the case, but other risks and benefits need to be closely studied before we can say whether they are safer or not.

Cons:

* Warmth
* Difficult to clear if flooded or leaking
* Uses significant gas to clear
* Difficult fit
* Needs third-party adapters to connect to loop
* Lost range of motion and peripheral vision
* Carbon dioxide dead space
* If mask fails, breathing loop fails
* Still need to carry backup mask and off-board second stage

Pros:

✓ Ability to record audio or talk through comms
✓ Warmth and environmental protection
✓ Potential to save an unconscious diver

Although my list of cons appears lengthy, I am not discounting the use of FFMs when needed. I found the Dräger Panorama/Nova mask to fit me very well and offer reasonable comfort, but it is not my choice to use a FFM mask routinely.If you choose to use one make sure it has a bite mouthpiece that will help prevent CO_2 buildup in the mask.

At the recent Rebreather Forum 3.0, the delegation unanimously agreed that further study off FFMs was warranted by the industry.

CHAPTER 3 - EQUIPMENT SELECTION

Weight Systems

Weight and trim can be tricky. Your instructor will work with you in a pool or confined water environment to get it just right. Weight is distributed in a variety of ways. Weights may be secured using several techniques:

Trim weight- secured to the top of the CCR to tilt the body into a comfortable swimming position.

Tank weights- some divers secure special third-party pockets to the tops of their onboard tanks to bring their head down into a comfortable swimming position.

Harness weight- trim weights may be attached to the harness shoulders to counteract buoyancy of counterlungs and loop. Ballast pockets or ditchable weight pockets are available on some rigs.

Counterlung weights- some rebreather counterlungs contain concealed pockets for lead.

Weight belts- weight belts and harnesses may be worn by divers wearing cold water dry suits that have significant positive buoyancy.

Trim weights are normally very small increments of one to two pounds. They

If you feel like your feet are dragging, you can weave trim weights into the shoulder harness to bring the top part of your body down and your feet up (above). Some units are equipped with ditchable weight pockets (below).

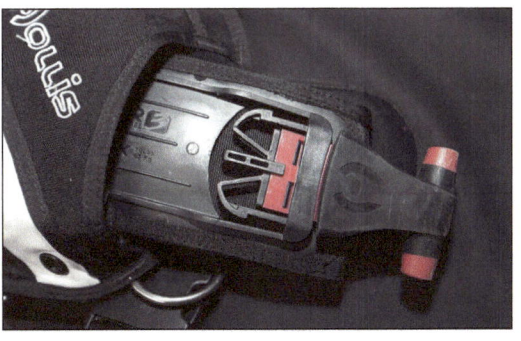

should not be too heavy, since they are generally not ditchable. They are only used to alter trim. If heavier weights are needed, they should be used with a quick release system on a belt or ditchable pockets.

Quick Buoyancy Tips

If you are struggling with buoyancy and trim on your rebreather, there are a few things you can consider:

If your feet are heavy, get lighter fins. *Heavy fins, such as the popular Jet Fin design, were originally designed for divers wearing thick neoprene*

CHAPTER 3 - EQUIPMENT SELECTION

drysuits. Current membrane-style dry suits rarely require negative fins unless you are diving with thick buoyant undergarments. Try leg gaitors to keep air out of your feet and consider a dry suit that is equipped with separate boots.

Get a rebreather specific wing. Not all rebreathers come with a harness and wing. There are wings that are specially designed to deal with trim issues faced by rebreather divers, optimizing the buoyancy cells lower towards the hips.

Snug up your counterlungs. The closer your counterlungs are to matching the anatomical position of your own lungs, the better they will breathe. If they are snugged up tight, then trim changes are minimized as you shift position in the water column. Think of them as a physiological extension of your own body. If they are loose and flop around, your buoyancy will shift with the air movement.

Custom-sized loop ring weights can counterweight a buoyant breathing loop. Solder or wire can also be wrapped around a loop hose to help hold it down.

Go to the dentist. The next time you are in the dentist's chair, ask her for her retired lead aprons. This convenient material can be cut into small trim weights or rolled into tiny packages that can adjust your trim. It can even be sewn into your dry suit underwear in the shoulder region if that is where the lead is needed.

Sheet lead. I once tried heating lead in a cast iron pan in an effort to pour my own custom weights. I probably added to my future dementia in the process. Now I purchase sheet lead from McMaster Carr or other suppliers. This lead is thin enough to cut with scissors and can be shaped into custom pieces or wrapped around the top of a small onboard cylinder. Team up with friends and buy a roll to share.

Shot pockets. Several manufacturers produce variable ballast pockets. These simple grommeted sleeves will hold 2 to 4 pounds of lead shot, which can be purchased at Walmart in the hunting section.

Custom trim weights. Many online dive shops carry custom lead or steel rebreather weights that are designed to fit specific rebreathers. This is likely the most expensive option, but it looks clean and well trimmed.

CHAPTER 3 - EQUIPMENT SELECTION

Consumables

Carbon Dioxide Absorbent

Every rebreather uses some sort of carbon dioxide absorbent material in the scrubber canister/filter. The job of this material is to remove carbon dioxide from the breathing loop. The material looks somewhat like cat litter. Some pre-packed sorb canisters look more like rolled white corrugated cardboard. There are many different brands of diving sorb and you should always use one that is recommended by the manufacturer of your rebreather. Rebreathers are designed and tested using a particular brand and type(s) of sorb. They may not perform properly if you use the wrong type. Medical sorb used in anesthesia machines is also made to a different specification and is not suitable for diving. It may look identical but it does not behave the same due to a different moisture content and other factors.

Once you have packed a scrubber canister with granular sorb, you should immediately install it in your rebreather as a part of your pre-dive check regimen.

Sodalime absorbent was first used in animal experiments as far back as the late 1700s, but it was not until World War I that scientists were rushed to develop reliable absorbents to be used in gas masks for soldiers. Today, the material is used in anesthesia machines, mine rescue equipment, fire safety, space technology, submarines and recompression chambers.

Within the human body, life sustaining energy arises from oxidative metabolism of stored food at the cellular level. Carbon dioxide is generated at a rate that generally matches the body's consumption of oxygen. This carbon dioxide is expelled from the lungs and exhaled into the breathing loop and soon reaches the scrubber canister/filter. The carbon dioxide passes through the absorbent material at a carefully engineered rate so that it has sufficient time to dwell in contact with the granules. A chemical reaction occurs along a "reaction zone" or "front" of unexpended material. The reaction zone slowly moves through the canister and converts the soda lime to a new chemical.

Sodalime is actually a mixtures of chemicals. The main components are calcium hydroxide [$Ca(OH)_2$], water, sodium hydroxide [$NaOH$] and potassium hydroxide [KOH]. The water in the granules facilitates a reaction with the carbon

CHAPTER 3 - EQUIPMENT SELECTION

dioxide that creates a carbonic acid. The carbonic acid reacts with the hydroxides to form soluble salts of both potassium and sodium carbonate. The soluble salts react with the calcium hydroxide and creates calcium carbonate. So, once the material is completely expended, you have essentially created limestone. When the materials is still active, it requires some special handling. If mixed with water, it creates a strong base that can cause alkaline burns. If flooded inside a rebreather, the diver can be endangered by inhaling or contacting the wet materials with their mouth or airway. This is referred to as a "caustic cocktail." Though rare, it is important to have safety protocols in place for dealing with such a situation. If you ever taste a metallic taste or feel a burning or soapy slime in your mouth, get off the loop, rinse your mouth with water and bail to open circuit for ascent. When handling and packing the raw material, you must avoid contact with your eyes, airway and skin.

Packing a scrubber canister takes patience. You should carefully follow the manufacturer's guidelines for your particular scrubber or filter.

Granular Sorb

The most common type of sorb is the granular form of soda lime. It is grayish white in color and is constructed of engineered grains or small tablet-like spheres. The size of the granule is generally described in the specification and it pertains to the size of a mesh that the grains will fit through. The manufacturer of your rebreather will not just recommend the brand of sorb, but also the granule size. Common names include Sofnolime and Sodasorb.

Micropore Canisters/ExtendAir

Micropore Inc. developed a Reactive Plastic Curtain (RPC) for use by the U.S. Navy in submarines and other enclosed environments. This material has been wound into a tight tubular canister for use in some specific brands of rebreathers. It is commonly referred to as ExtendAir. These cartridges come in different sizes,

CHAPTER 3 - EQUIPMENT SELECTION

that are designed to fit specific models of rebreathers. Ensure you get the correct size for your unit.

Storing sorb properly is important too. Follow the guidelines in the Material Safety Data Sheet (MSDS) for information, but in general, it should be stored above freezing. Freezing can damage and shatter the granules, creating dust that can be problematic in the breathing loop and during packing. Avoid any activity that can damage or crush particles of absorbent material.

Canister Design

The material inside a scrubber does not behave the same way in all canister designs and all rebreathers. A common misconception is that five pounds of absorbent will net the same canister duration in any rebreather. This could not be further from the truth. The canister vessel is specifically designed to optimize the life of sorb in a particular rebreather design, but many factors affect the life of the sorb. Scrubber efficiency is affected by temperature, depth (relative gas density), granule size and the length of time the gas is in contact with the absorbent material (dwell time). The temperature of the gas inside any given model of rebreather can vary greatly depending on how well the unit is insulated and other factors such as canister orientation. The dwell time of the gas within the scrubber material can be affected by design elements that either speed up gas velocity or create restrictions that slow it down and force it to remain in contact with material longer. The volume of scrubber material versus your breathing volume will also affect duration. Common designs include axial, radial and crossflow canisters.

What do I do with my sorb when I'm finished with it?

Modern rebreather absorbents are made of sodium hydroxide and calcium hydroxide primarily. Once the exothermic reaction is complete, the by-products of water and heat are released and you have essentially created chalk or lime. I use this material to fill the gaps in my limestone driveway.

If you are anywhere other than your own property, ensure you follow the local guidelines for disposal. Some parks provide specially-marked garbage cans for expended sorb. Some boats will not allow packing or emptying scrubbers on their vessel.

CHAPTER 3 - EQUIPMENT SELECTION

Batteries

Electronic rebreathers are operated with either disposable or rechargeable batteries. The electronic system usually monitors the voltage of batteries and often provides a warning system when batteries are low. Checking the voltage of batteries is a part of a pre-dive check. The manufacturer will give you specifications on when batteries should be replaced.

Many batteries in use these days are Lithium Polymer or Lithium Ion construction. I learned about the volatility of lithium batteries twice in my rebreather career. My first rebreather contained a set of handmade lithium batteries that powered the unit. These battery sets cost about $85 each and ran for around 40 hours. We weren't sure how long they would last and were keeping careful performance logs. We were pushing the limits of their life to track duration and see how the computer would behave as they were being depleted. On a dive in a silty cave at close to 300 feet of depth, I heard a loud banging that sounded like somebody had smashed a tank against the wall with all their strength. Simultaneously, the solenoid on my rebreather began firing erratically, sending oxygen into the loop when it was definitely not needed! I shut down the oxygen injection and switched to operate the rebreather manually, aborting the dive. Surfacing after about an hour of decompression, I discovered that there was a large hole in the back of the rebreather. A C-cell battery had actually exploded. I learned two things that day. First, lithium batteries that are deeply discharged or damaged by impact are volatile and may explode. Second, I never want a battery anywhere inside my breathing loop. Some rebreather designs would have left me choking on the result of a battery explosion. Luckily, my unit was designed with the batteries external to the breathing loop. I was able to abort the dive safely and in an unremarkable fashion, only because the designer knew it was not a good idea to put batteries inside the breathing loop.

Cave explorer Dr. Richard "Harry" Harris knows how hot and fast a fire will burn when ignited by lithium batteries. Little was left of his car (above) or rebreather (below) after a scooter battery caught fire during an expedition in Australia. Photos: Richard Harris.

CHAPTER 3 - EQUIPMENT SELECTION

The second time I was reminded of the volatility of lithium batteries was when an underwater cinema light exploded on my porch while it was being charged. Luckily, it was contained in a ventilated charging box, otherwise it may have burned down my house. This battery technology is still rather young. Read the manufacturer's warnings and charge batteries according to their specifications. Consider *where* you are charging them and only do it when you are home and able to respond to an explosion or fire. Consider replacing rechargeable batteries early since older batteries tend to become more volatile.

Gas

Your rebreather will carry one or two tanks onboard and one or more bailout tanks that need to be filled regularly. You'll be getting air fills, high PO_2 nitrox and possibly oxygen fills, depending on the model of rebreather. If you don't already have it, you will receive training in the handling of rich gases. High pressure oxygen, when mishandled, can cause fire and explosion. Ensure you understand the risks and how to mitigate them.

> ***How much bailout gas do you suggest carrying on a dive?***
>
> *People have very differing opinions on the topic of bailout. I prefer to carry adequate personal bailout gas to get me to the surface safely with a solid margin for error. When extremely long-range dives warrant carrying more than two 80cft tanks in addition to my rebreather, then I usually stage gas in depots within the cave environment. Deco gas that is staged in the cave is sometimes planned as shared gas since both my partner and I are unlikely to need open circuit deco gas at the same time. On a deep ocean dive, I will carry two to four tanks personally, but must rely on a support team for anything beyond that scope. I approach each dive as an independent risk assessment and try to make the best choices with extreme conservatism.*

Sensors

Oxygen sensors are an additional type of consumable that you will use in a rebreather. As mentioned earlier, they have a shelf life and must be proactively changed to prevent feeding false information into the rebreather electronics system. You'll need to track their life span and change them out every 12 to 18 months or sooner. The manufacturer will recommend an interval for replacement. When I replace a sensor, I put a note in my SmartPhone calendar on a date one year ahead and 18 months ahead. It simply says "Sensor One- 12 months old." That gives me time to order a replacement and install it. At 18 months, the note says, "Sensor One expired." I always label a sensor with an "installed date" and position in the rebreather; 1, 2 or 3. That seems to prevent mixups.

CHAPTER 3 - EQUIPMENT SELECTION

Most rebreather manufacturers will automatically replace your sensors if you send your unit for service and if sensors are aged beyond their recommendations. That might be a 12- or 18-month replacement interval. Ensure you know the age of your sensors and whether they are being swapped out during routine service or whether you need to do it yourself.

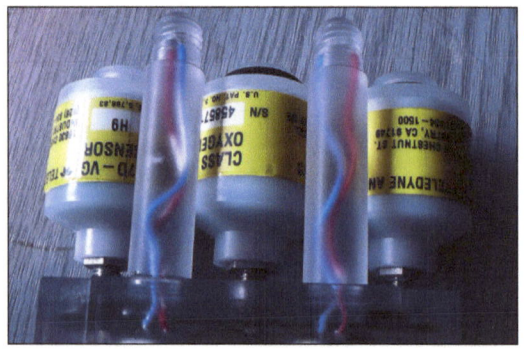

The sensor on the left, labelled "H9" was manufactured in August 2009.

Golden Rules For Sensors[3]

1. Change oxygen sensors every 12 to18 calendar months once opened.
2. Do not use sensors that have been stored for over two years.
3. Do not use medical, industrial or automotive sensors in a rebreather.
4. Only use a sensor designed and tested for use with your rebreather.
5. Do not store sensors in air tight bags.
6. Do not store sensors in high PO_2 for extended periods (+24 hours).
7. Do not store sensors in high temps (+50 degrees C) or in dry gas.
8. Do not store sensors in an inert gas.
9. If sensors do not give similar readings on a dive, then do a diluent flush to confirm the readings.
10. To test for current limiting (see page 139 for detailed discussion), occasionally do an oxygen flush at 20 feet/6m or a diluent flush at depth.
11. Only ever shake moisture from cell membranes or carefully use tissue paper, do not use tools/cloths.
12. Keep electrical connections clean.
13. Do not block the holes on a sensor's body.
14. Allow the rebreather's temperature to stabilize before calibrating (if it has been in sunlight/cold and has heated up/frozen). Better, calibrate when the absorbent canister has reached its operating temperature.
15. If you get erratic sensor readings (in known gas content) the sensors may be damaged and should be replaced.
16. Replace sensors that have been flooded.
17. Never swap a sensor to another position in the electronics without recalibration.

CHAPTER 3 - EQUIPMENT SELECTION

Budgeting for Consumables

Here is a breakdown of consumables costs based on an average rebreather dive. Many variables exist for particular units and tank sizes.

Oxygen- in North Florida a rebreather tank can be filled for about $4 to $6 and lasts up to ten hours, independent of depth, but don't think about starting a dive with a half-filled tank.

Diluent- depth and profile dependent, but lasts for a good day of diving for $4 to $6, but don't start a dive with a half-filled tank.

Bailout- these are likely plumbed to your inflator and/or dry suit, but won't be used for anything else beyond safety checks, practice or open circuit bailout, therefore will last many dives prior to refilling unless you have had an incident that forced you to use open circuit.

Batteries- depends on your unit. Some are rechargeable and therefore extremely low cost.

Sorb- averages between $20 and $30 per fill or cartridge for most units. These may last from two to six hours depending on your unit, although it is important to start fresh for deep dives.

Trimix- local prices vary dramatically, but you will likely only need roughly 13 cft (2 liters) of mix for a deep dive. If you keep a filled pair of bailout tanks, you'll only be refilling the onboard tank for dives. It is cheaper to store an extra set of deep bailout tanks than fill and dump gas on a regular basis.

Service- don't forget to factor service costs (and shipping for service) into your plans. Don't wait for something to break. Be proactive.

Basic Rebreather Care

After each day of diving, your rebreather will need some essential care. Your rebreather should come with a post-dive checklist and/or cleaning protocol that instructs you specifically what to do if you will be diving the following day or if you are putting the rebreather away for a longer time period. As soon as possible after diving, you will disassemble the breathing loop and sanitize it to

CHAPTER 3 - EQUIPMENT SELECTION

prevent the growth of undesirable bacteria or pathogens. A variety of products may be recommended including Steramine, Virkon, Betadine or other disinfectants. Follow the manufacturer's guidelines. Your loop is an extension of your lungs. Anything that grows in a moist environment can be introduced to your lungs during diving.

Rebreathers, like regular SCUBA gear, need periodic service. A maintenance schedule should be included in your user manual. Some items may be replaceable by the user and a schedule will inform you about when you need to have the unit serviced at a dive shop or service center. Rebreathers have unique service requirements and components should only be serviced by qualified and factory-authorized technicians.

"Discover Rebreather" pool demonstrations should be equipped with a way to sanitize the rebreather loop between divers.

Cleaning Electrical Contacts

Always use cleaners specifically designed for the precious metal contacts used in low current applications found in high-tech electronics. The low current makes the contacts exceptionally sensitive to contamination yet their ultra thin precious metal plating is easy to damage during cleaning.

- ✓ Do not use standard SCUBA cleaning solvents.
- ✓ Do not use rubbing alcohol or tuner cleaner which may contain petroleum distillates.
- ✓ Do not put silicone grease on electrical contacts.
- ✓ Do not use a pencil eraser on precious metal contacts such as gold. The eraser is abrasive enough to remove the thin gold plating.
- ✓ DeoxIT Gold GN5 Contact Cleaner is excellent for maintaining electrical contacts on your rebreather or diving computer. It is suitable for delicate precious metal contacts. It is also non-toxic and environmentally safe. This product comes in a small spray bottle and is also considered safe if it spills onto plastic materials.

Tools

There may be a few tools that come with your rebreather, but you should also compile some additional specialty items to make handling your unit easier. Some of these items might include:

- ✓ oxygen analyzer, marking tape and pen
- ✓ clipboard or computer/iPad application with checklists

CHAPTER 3 - EQUIPMENT SELECTION

- ✓ lubricant (oxygen compatible and silicone if approved, such as Tribolube 71, Christolube MCG 111 and Dow Corning 7 Silicone)
- ✓ O-ring removal tools (plastic and brass)
- ✓ toothbrush for cleaning small parts
- ✓ a rubber mat for protecting the base of your scrubber when filling
- ✓ a pouring cup for sorb
- ✓ disinfectant tablets
- ✓ a rinse/soaking bucket for cleaning rebreather parts
- ✓ vinegar for removing caked expended sorb (if recommended by manufacturer)
- ✓ a stiff plastic or nylon bristled brush for cleaning rebreather parts
- ✓ gloves, dust mask and safety glasses for packing sorb
- ✓ Deoxit Gold GN5 Contact Cleaner
- ✓ magnifying glass to examine small parts

The PRISM2 rebreather is designed to be cleaned up inside its own scrubber bucket. This bucket contains Steramine disinfectant that colors the water blue. It is a food grade product used in the restaurant business.

Spare Parts

Some manufacturers sell a "Save-a-Dive" kit for their rebreather. This kit contains O-rings, seals, small parts and sometimes sensors. I highly recommend that you get one of these since O-rings might be proprietary sizes, construction or material. You should also carry spare batteries if applicable and consider having spare sensors on hand if they are not easy to get locally. Remember, those sensors will age whether used or not, so you may end up disposing of unused sensors that sit around too long. Additionally, you can compile the types of spare parts you might have already been using for open circuit diving. Spare hoses, O-rings, HP spools, zip ties and mouthpieces are always useful. In fact, mouthpieces are more important that ever, because a hole in a mouthpiece can cause a catastrophic loop failure that allows water to flow into the breathing loop. I prefer using moldable Seacure mouthpieces since they can be custom fitted and rarely leak. Make sure they are secured well with a zip tie that is checked regularly. Though some people claim these mouthpieces make them gag, if you follow instructions, Seacure describes how they should be trimmed so they do not induce that sensation.

[3] Gurr, Kevin. "Oxygen sensors for use in rebreathers." - White paper written in 2013 and distributed on RebreatherPro.com and other outlets.

A diver leaps off the stern of the boat carrying double tanks and two stage bottles. Beside him, his partner wears a heavy rebreather and bailout tanks. A decade ago, technical diving was in its infancy, yet today, it is the fastest growing segment of the diving community. Just over 20 years ago, the Dive Equipment Manufacturer's Association (DEMA) Trade Show banned all vendors who had anything to do with Enriched Air Nitrox, yet now, we are in a space-race of new technology. Given that the once maligned sport of rebreather diving is rapidly emerging as the most exciting opportunity in the underwater world, there are a few things to remember in regards to taking responsibility for yourself and your activities.

Interview Your Prospective Instructor

It is critical to seek out proper training from a qualified and experienced instructor. Your instructor should have plenty of real world diving experience, (outside of teaching) and be willing to share their background with you.

Retain Control

During training you will be significantly expanding your horizons. Your instructor should act as a mentor but ultimately you should always be capable of self rescue and buddy rescue at any time. If you feel like you are out of your league, then you may be moving too fast.

Use Gear That Works

Invest in reliable equipment and have it serviced properly and regularly. This isn't the time for frugality. Your equipment is life support. Don't enter the water with inadequate life support or anything that is not working in top order. There is no such thing as a simple dive.

Be Extra Cautious on Guided Dive Experiences

Be wary of guided adventure diving that takes you beyond your training and background. A guided cave dive without proper training could get you killed. Similarly, deep bounce dives might take you beyond your experience level into dangerous territory without adequate backup.

Don't Trust Anyone

Buddy diving is always preferred over solo diving, but be careful not to surrender too much control to your diving partner. You should always be capable of getting yourself back to the surface safely without their aid if necessary.

Don't Ignore Safety Issues

Every dive should begin with a thorough pre-dive safety check of equipment. In the case of rebreather diving, that will also include a five-minute pre-breathe of the unit itself. If you find an issue during your pre-dive check, take the time to resolve the problem or abort the dive completely. No matter how far you have traveled and no matter how much money you have invested, there is no dive worth the risk of using damaged equipment.

4

Training

In this chapter:

- *How We Learn*
- *Training Programs*
- *Choosing a Training Agency*
- *Finding the Right Instructor*
- *Preparing for Class*
- *Class Content*
- *Staying Current*
- *Specialty Training*

How We Learn

Technical diving and specifically, rebreather diving, is a continual learning process. If we closely examine how we learn, we can better prepare for the pitfalls associated with each stage of the learning process.

Gordon Training International is popularly considered to be the originator of the "Conscious Competence" model, which describes the steps of learning any new skill. This model is particularly applicable to rebreather diving.

The model describes the first stage of learning as "unconscious/incompetent" or "unconscious/unskilled." This stage describes a rebreather diver on his or her first day of class; you are unaware of the proper function of the unit and incapable of determining risk. You simply don't know what can kill you. It's like your first day learning to drive a car. You're like an invincible sixteen year old with a deadly weapon.

The second stage and each step thereafter is often associated with a sensation of awakening, when you feel *"like a light bulb turned on."* As you make this step forward, you enter the realm of "conscious/incompetence." At this point, you are beginning to understand the function of your unit and are able to assess risks, but still need close supervision. If you were driving a car, this would be the point where you are able to move it down the road effectively and are somewhat aware of the surrounding traffic issues, but you are still making wide turns and are incapable of parallel parking. At this stage, your mom is still sitting in the passenger seat. As a student diver, your instructor is watching you closely while you gain experience and learn new skills.

CHAPTER 4 - TRAINING

Next, the learner reaches the point of "conscious/competence." This may be the point when you complete your initial rebreather training program. At this level, you have mastered basic controls, have a good assessment of risk and are able to complete self- or buddy-rescue. This may indeed be the point where you are the safest rebreather diver you will ever be. You still have a healthy fear that the unit may fail and are consciously piloting the rebreather with great care. If you were driving a car, this is the point where you would have earned your license. You are likely worried about damaging Dad's car and are driving with your hands in the correct position on the wheel. Hopefully, you are not texting or eating Big Macs as you head down the road!

The final stage of learning occurs when you reach the "unconscious/competent" level. This is akin to someone who has been driving a car for a long time. When you make a daily commute, sometimes you barely recall the route you took or the things you saw along the way. When this occurs in rebreather diving, it is often the point when complacency kicks in. Rebreathers are well-made and very reliable. When you log a series of incident-free dives, you may start to get casual about how to treat the unit. You may think that nothing can ever happen.

Rebreather divers with roughly 50 to 100 hours after their initial training, may be at the greatest risk, especially if nothing has scared them along the way. A serious malfunction during that timeframe often frightens you back to the previous level of learning, where you become a conscious driver of your unit again. A long absence from diving will also result in stepping backwards in the model until you catch up with skills and practice.

To avoid the pitfalls of complacency, good procedure and a commitment to pre-dive checklists and proper pre-breathe sequences are critical. A diver who carefully reviews their personal preparedness as well as their equipment readiness will be better prepared to deal with the issues on the road ahead. This takes concerted effort and uncompromising procedures but is essential for your longterm enjoyment and survival in this sport.

CHAPTER 4 - TRAINING

Training Programs

In such a new form of diving, there is still a lot of bantering over class nomenclature and even training standards. It is worth a little primer about who is involved in writing training standards.

In the United States, we refer to CCR diving hobbyists as "sport rebreather" divers. That includes both recreational and technical markets. In a nutshell, if you are not being paid to dive for a job and it is something you do for fun, it is considered "sport diving." Commercial diving falls under strict guidelines from OSHA. Manufacturers, operators and training agencies self-police their ranks to minimize accidents and deal with safety issues in an effort to keep sport diving out of the realm of governmental rules and regulations. This means that our sport is always evolving. As we learn more from accident analysis, the generally accepted rules within our ranks will continue to evolve. Classes will reflect new revelations learned as our sport expands.

In the dawn of sport rebreather diving, manufacturers decided to stay out of the business of training. Training agencies such as the International Association of Nitrox and Technical Divers (IANTD) possessed a captive international audience of divers that were clambering for information on new technology. It made sense for manufacturers to concentrate on design and testing, and training agencies to develop educational programs. It was not immediately clear whether training would be specific to individual units or generalized. As more training agencies joined in the development of programs, it seemed clear that general knowledge and safety protocols were important, but so were skills that were specific to individual units. This created an even greater need for manufacturers and agencies to work together.

In 2010, several manufacturers formed an organization called RESA (Rebreather Education and Safety Association). This organization was created to promote standards-based manufacturing of rebreather equipment and to promote a high level of safety awareness and training. It was also designed to react, as a group, to industry wide issues: to date RESA has agreed to an absolute minimum level of third-party product testing, to make product testing protocols readily available to equipment inspectors following a diving incident, encouraging manufacturer involvement in dive incident analysis, to promote the use of checklists and to improve pre-dive preparation. Recently RESA has begun to suggest minimum standards for entry level through instructor training. As a result, training agencies are being fluid in their efforts to participate with RESA, to understand

CHAPTER 4 - TRAINING

and implement changes that reflect RESA's leadership. Training standards are revised by agencies annually and are growing to reflect an emerging global consensus on issues that build a safer sport for all.

The largest international training agencies involved in rebreather programs are: PADI (Professional Association of Diving Instructors), IANTD, TDI/SDI (Technical Diving International), IART (International Association of Rebreather Trainers), NAUI (National Association of Underwater Instructors), RAID (Rebreather Association of International Divers), PSAI (Professional SCUBA Association International), SSI (SCUBA Schools International) and ANDI (American Nitrox Divers International).

PADI has created a delineation between recreational and technical rebreather equipment and training. They devised a classification that refers to individual rebreathers and training as either "Type R" or "Type T" for recreational and technical. Their standard describes the differences between rebreather types:

Type R Rebreather:

A Type R Rebreather is an eCCR or eSCR intended for divers making recreational, no stop (no decompression) dives. It will normally have the following characteristics (among others):

- pre-packed or easily packed scrubber canisters
- will not operate or will warn the diver if the canister is missing
- has a system for estimating scrubber duration
- has warnings for low or closed gas supply, low battery life, high or low PO$_2$
- open-circuit second stage supplied by diluent cylinder for sharing gas
- "black box" data recorder function
- automatic setpoint control (eCCR)
- HUD (heads up display) warning system in the diver's line-of-sight during normal diving
- BOV operable with one hand
- electronic prompts for pre-dive check

Type R rebreathers are intended for recreational divers; divers can qualify to use a Type R rebreather in the PADI Rebreather Diver and Advanced Rebreather Diver courses.

CHAPTER 4 - TRAINING

Type T CCR:

A Type T CCR is an eCCR or mCCR intended for divers making technical deep decompression and cave dives. Type T characteristics include:

- may have user-packed scrubber system

- warnings for low battery life, high or low PO_2

- "black box" data recorder function or used with a computer that serves that function

- automatic setpoint or manual setpoint control capability (if eCCR)

- HUD (heads up display) warning system in the diver's line-of-sight during normal diving

- BOV operable with one hand (Note: If you will dive in cool water with insulating gloves, be sure you can still operate the BOV with one hand. Some models have optional cold water attachments for this purpose.)

- must have manual diluent and oxygen valves

- functions to a depth of 100 metres/330 feet (with appropriate diluent and secondary life support systems)

- must have isolated secondary electronics for operational confirmation and manual operation by user in the event of a primary systems failure

Type T CCRs are intended for technical divers; you will qualify to use a Type T CCR in the PADI Tec 40 CCR Diver course and higher levels.[4]

Choosing a Training Agency

The base of rebreather instructors around the world is still relatively small and training on some rebreather brands may not be offered by all training agencies. You may find that your decision regarding which training agency to select is based on finding an instructor first. Some training agencies offer comprehensive online training and testing prior to signing on with an open water instructor. Others offer good general training materials that may be useful no matter which agency you are ultimately qualified under. In my opinion, it is more important to seek out the right instructor first.

CHAPTER 4 - TRAINING

Finding the Right Instructor

Choosing a rebreather is tough, but finding the right instructor might be even more difficult. With new models hitting the market on a regular basis, how do we define experience? Does your instructor actually dive their rebreather in conditions similar to those in which you intend to dive, or do they spend all their hours teaching in controlled scenarios? Rebreathers are used in a variety of environments for scores of different purposes. As a result, certain rebreather instructors specialize in certain types of diving.

In 2005, University of Hawaii, Bishops Museum Ichthyologist, Dr. Richard Pyle offered some excellent thoughts on gauging experience. A CCR diver since 1994, he has "put some miles on the tires" of his MK-5P rebreather. "After my first 10 hours on a rebreather, I was a real expert. Another 40 hours of dive time later, I considered myself a novice. When I had completed about 100 hours of rebreather diving, I realized I was only just a beginner. Now that I have spent more than 200 hours diving with a closed-circuit system, it is clear that I am still a rebreather weenie." Consider that an instructor may be qualified after 50 to 100 hours on a new unit, depending on their training agency. As a potential client, you should query your instructor's background and logged hours to determine if their experience matches your expectations. Most of us that have been diving CCRs for long, become more humble as we accumulate our rebreather diving hours.

There are a few things you want to know about your instructor:

✓ Do they own their rebreather and dive it often? Do they teach many different rebreathers or are they a true specialist in one brand?

✓ Do they maintain their equipment professionally? Are there divers that can attest to their attention to detail?

✓ Do they have real-world experiences outside of teaching? Are those environments similar to ones that you will be diving in?

✓ Do they have a current professional rating and insurance?

✓ Do they charge a professional fee that supports their investment in gear/training?

✓ Are they on the manufacturer's recommended list?

✓ Are they considered to be a good role model in the diving community?

When you are eventually certified to dive on a rebreather, you are qualified to dive in an environment that is similar to the training environment. If you flew down to Florida's clear open water to take a class and are now planning to head out on a cold water wreck dive, then there will be some need for post-class baby steps. You cannot expect to jump right back in to the advanced level of open circuit diving you were doing before the class. If you have an opportunity to train in an environment that is as challenging as your home diving environment, then that is ideal.

CHAPTER 4 - TRAINING

Preparing for Class

If you purchased a new rebreather, it is customary that the manufacturer has shipped your rig to your instructor and you haven't even had a chance to get acquainted with it. Contact your instructor to ensure that you know what auxiliary gear will be necessary to complete your class. This might include a wing and backplate if your rebreather does not come with one and will also include items such as a bailout tank and regulator.

If you have prepared yourself academically prior to class, then you will have more time for hands on practice and in-water training. Kristine Rae Olmsted (below) prepares herself for diving fitness through a CrossFit program at her local gym. Photo courtesy of Hannah Nowill, HannahandRandall.com.

The first day in the water can be very frustrating, especially for highly experienced divers. People with excellent skills find themselves reduced to the level of neophyte in new equipment. Just remember, if you weren't going to make mistakes, then you wouldn't need the guidance of an instructor! Your instructor knows that you will be a far better diver as soon as she is not looking at you! Nothing can adequately prepare you for the new experience of buoyancy on a rebreather. All your years of perfecting breath-controlled buoyancy will be out the window. When you exhale into a rebreather, the gas stays in the loop. That means your buoyancy is static. You won't drop. It will feel really strange at first, but you will quickly learn new strategies for buoyancy.

You can prepare ahead for the academic work. You should be very comfortable with the concept of partial pressures and be able to quickly calculate the partial pressure of a gas at a given depth... *preferably while you are in the water.* Refresh the knowledge and skills from the Advanced Nitrox level if applicable, and be very comfortable with task loading. Are you comfortable diving without a mask? Shooting a lift bag? You will be asked to perform these skills wearing the rebreather, so ensure your general competence on open circuit.

CHAPTER 4 - TRAINING

Most importantly, be physically and mentally well. Don't plan on late nights visiting friends or partying. A rebreather class is intense and draining. You need to be sharp and well rested. The rebreather class is also very experiential. Academic concepts and skills all come together through repetitive, in-water training. What seems a little fuzzy in a textbook, will make sense once your instructor has demonstrated a skill and you have had the chance to practice it. There is nothing more important than plenty of time in the water and lots of repetition of skills. So, pack your sense of humor and patience. It may be awkward at first, but it is the beginning of an exciting new world of diving.

Class Content

Whether you enroll in a recreational rebreather class or an introductory technical CCR class, there are some basics that you will cover. To begin, you will complete a series of liability releases, assumption of risk forms, medical questionnaires, insurance information forms and other paperwork. Most CCR instructors will require that a physician reviews and approves your medical history prior to any in-water training, so ensure you have time to get that done before class.

The first part of class includes a little home study. Even if your rebreather is sent to your instructor, you should review the owner's manual before class. These are generally available on the manufacturer's website. You should also find out what student materials are needed for class and review them in advance. There may be homework requirements and online testing prior to meeting with your instructor. You will be making an investment in good training, so ensure you arrive at class prepared and with your reading completed. If you don't, you will likely lose precious water time that was planned for you. Each training agency requires that divers master a certain skill list and generally requires a certain number of in-water hours or dives, but your instructor has also likely left a significant buffer for time in the water. You'll want to be able to take advantage of every moment of water time you can get your hands on. Time equals experience.

You will participate in a number of hands-on activities with your instructor prior to diving. These activities include component inspection, building the rebreather, navigating the handset, using pre-dive checklists, conducting a pre-breathe, setting up equipment, fit issues (counterlungs, OPV), post-dive procedures, maintenance demonstrations, oxygen cell calibration, dive planning and setting up the onboard and offboard computer(s).

CHAPTER 4 - TRAINING

Your rebreather class will then move to a confined water environment, which might be a pool or a controlled open water setting where you can see clearly underwater and easily stand up. In this environment, the instructor will walk you through the basics and demonstrate skills, allowing you ample time for practice and mastery. Only once those skills are demonstrated, reviewed and mastered, you will move into a more challenging open water environment. You'll begin shallow and ease your way into longer and deeper dives throughout the class.

Your instructor will use a pool or confined water area to demonstrate and give you a chance to master general skills before taking you to an open water diving site that has additional environmental challenges.

Each brand of rebreather has a specific set of skill requirements that must be met prior to earning your certification, but in general, you will learn the following skills in a recreational rebreather class:

- ✓ Assembly, pre-dive check and pre-breathe sequence
- ✓ Proper fit and adjustment of rebreather
- ✓ Weighting and trim
- ✓ In-water checks (leak check, buddy check, bailout check)
- ✓ Donning bailout tanks (if applicable)
- ✓ Controlled descent
- ✓ Controlled ascent
- ✓ Using the BOV, loop protocols
- ✓ Clearing a flooded mask
- ✓ Operating the BCD/wing (addition and dumping), inflating the wing (oral and manual)
- ✓ Operating the rebreather with a minimal loop volume
- ✓ Swimming with a minimal loop volume
- ✓ Removing water from the loop
- ✓ Using and monitoring the onboard computer
- ✓ Using the HUD and secondary display
- ✓ Using offboard gas for bailout
- ✓ Helping another diver
- ✓ Buoyancy control and hovering
- ✓ Making safety stops
- ✓ Procedures for omitted decompression
- ✓ Using DSMBs
- ✓ Using a drysuit with your rebreather (if applicable)
- ✓ Safe entries and exits from a boat or shore
- ✓ Emergency bailout procedures
- ✓ Open circuit ascents
- ✓ Post-dive procedures

75

CHAPTER 4 - TRAINING

In addition to skills listed in the recreational class, you will learn the following skills in a technically-oriented class:

- ✓ Buoyancy and trim adjustment with additional bailout
- ✓ Manual operation (static, swimming, ascending)
- ✓ Sensor confirmation (diluent flush)
- ✓ Sensor confirmation (checking for current limitation)
- ✓ Switching to open circuit, verification and problem solving and switching back
- ✓ Identifying issues underwater
- ✓ Hypoxia drills
- ✓ Hyperoxia drills
- ✓ Hypercapnea Drills
- ✓ Switching setpoint underwater
- ✓ Partial flood recovery (static and swimming)
- ✓ Use of DSMBs in decompression
- ✓ Automatic operation (static, swimming, ascending)
- ✓ Dealing with gas emergencies
- ✓ Disabling cells (if applicable)
- ✓ Operating in SCR mode
- ✓ Ascending in SCR mode (if applicable)
- ✓ Removal and replacement of bailout tank (static and swimming)
- ✓ Dealing with various alarm scenarios
- ✓ Decompression skills (if applicable)
- ✓ Removal and replacement of unit on surface
- ✓ Removal and replacement of weights on surface
- ✓ Maintaining PO_2 at stops
- ✓ Complex emergency drills
- ✓ Rescue
- ✓ General diving skills review
- ✓ Other brand-specific skills

Staying Current

I recently returned from riding my bicycle across Canada on a promotional effort for my film *We Are Water*. That meant that I was dry for four months. In my quarter-century plus diving career, I have never had such a long hiatus from diving. I knew that when I got home, I would need to take a lot of time to maintain my equipment and refresh my skills on each different rebreather and open circuit configuration. Some equipment, left sitting for so long, needed lots of attention. Lubricating grease thickened and stiffened parts. Sensors had aged. Some gear walked easily through a functional assessment and was ready to go. Some equipment tested fine on the bench yet had issues after getting wet again. More importantly, my brain needed to be re-engaged.

I have the luxury of being able to jump into shallow open water across the street from my house, but you may need to find a local pool to get back up to speed. You can use the lists above to ensure all your skills are still mastered. I

CHAPTER 4 - TRAINING

searched the Internet to ensure there were no needed upgrades. I re-read my user manual and updated my checklists. I sent one unit overseas for service. Then I made a few easy dives and practiced my skills on each rebreather. It doesn't matter how much overall experience you have, if your motor skills are not fresh and your mind is not sharp, then you will forget things. Competency includes currency.

The Right Tool for the Job

Some rebreather divers fully commit themselves to CCR diving and choose to abandon open circuit technique and equipment. Personally, I believe that rebreathers are tools and as such are not always the right tool for the job. When I need to thread my body through delicate cave formations, it is definitely not the best tool. I can more naturally coast through precise buoyancy changes on a set of sidemount tanks. Sometimes the camera or scientific equipment that I am operating requires a great deal of focus and attention. In some of those cases, I feel that certain rebreathers require more attention than I can easily offer and prefer to stick to open circuit. Sometimes I cannot get the proper logistical support for rebreather. I also believe it is important for me to use the same gear configuration as my students. They need a role model to watch and learn from. That means that I do not use a rebreather if I am teaching an open circuit cave class.

Is a rebreather or sidemount SCUBA appropriate for this dive site? Choose the tool that you are proficient on, meets dive objectives and is appropriate for the environment. Lower photo of the author by: Gene Page.

The people that say, *"once on rebreather, always on rebreather,"* are not saying it is unsafe to dive open circuit. They are merely reminding you that proficiency and well practiced motor skills are important for safety. If you just made the switch back to your rebreather, it might take a bit of practice to get back in the groove.

77

CHAPTER 4 - TRAINING

Specialty Environments and Training

This book is primarily intended to answer all the questions of aspiring and new rebreather divers. However, you are probably already dreaming ahead about the diving you will be able to do with your new toy. Once you have completed your class it is important that you take time to gain experience. You can't jump back into the deep open circuit trimix dives you were doing right before class. You can't launch yourself into the back of the cave that you dive every weekend.

There are many levels of advanced rebreather training available when you are ready and have met the prerequisite hours which may be between 25 and 50 hours after initial training. The bottom line is that you need to be completely familiar and competent with your rebreather before you add a new and significant task load. Agencies differ greatly on how they handle increased depths and mixed gas training. You'll need to check with your prospective instructor to find out how you can progress. The qualified instructor pool gets smaller and smaller as you head up the ladder. Your instructor will not only need to have rebreather qualification and currency but they will also need to specialize in the environment that interests you. You may have to book several months to a year ahead of time to get the instructor of your choice.

If you plan on advancing in your specialty training you may need to look early for an instructor that is experienced in both CCR Trimix and the particular environment that you wish to train in.

Specialty diving such as CCR Cave diving is gaining popularity too. In this case you may enter a class as an experienced CCR diver with no cave diving background or you may start the program as an OC cave diver with a relatively new rebreather. The class will be tailored to your background and experience, eventually reaching the same result of creating a competent CCR Cave Diver.

CHAPTER 4 - TRAINING

Upgrades and Modifications

Your rebreather may need some upgrades prior to taking a class for diving beyond 130 feet/40m. Many recreational rebreathers are not suitable for deep diving at all. Some eCCRs are programmed to offer limited depth and air diluent for the first phase of training. After successful qualification and experience hours, you can apply to the manufacturer to unlock advanced features. This is often handled with a PIN code that is given to your instructor so they can upgrade your rebreather at the appropriate time. Some rebreathers are completely unlocked and rely on you to make judicious steps towards more advanced diving. Learn about the expectations of your manufacturer and training agency and plan ahead so you are not left without a PIN code on the day you start your class.

When you embark on deeper dives, you will need several bailout tanks, decompression tanks, offboard gas connections and other auxiliary equipment.

Some manufacturers offer advanced optional equipment for advanced and specialty diving such as offboard gas whips that accept connections from your offboard tanks. In some cases, the offboard tanks can be plumbed completely into the rebreather supplying gas to the solenoid, BOV, ADV, etc. In other cases, your model may not offer these features. This is where you enter a minefield of options. Third-party BOVs, offboard connectors, weights, stands, HUDs and fourth-cell monitors are all being used by divers in our community. Some work great and some are completely unvalidated. Be careful to do diligent research about whether any modifications affect your warranty, work of breathing, carbon dioxide dead spaces or other factors. A plumbing nightmare of connections may introduce issues that you never considered. Replacing the counterlung with a neoprene one may introduce a new valve that does not have the same flow restriction as the one that came with your unit. It might allow an

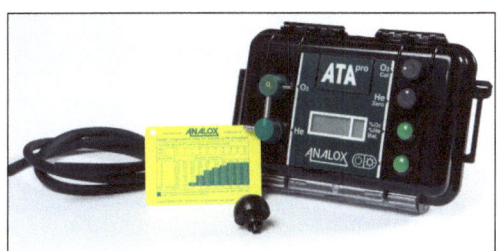

For mixed gas diving, you will likely need a helium analyzer unless you have one that is easily accessible at your local shop.

CHAPTER 4 - TRAINING

oxygen free flow to supply a dangerous injection of pure oxygen into your loop. A slider valve to cut off oxygen supply might be convenient, but what happens if it is accidentally shut off?

I am often asked about the modifications I have made to my rebreathers. I have owned many brands over the years. I would advise anyone to dive a rebreather straight out of the box as the manufacturer tested and intended it to be used. If you need to make a lot of third-party additions to your unit, you may be creating unintentional safety issues. You may also be taking the rebreather beyond its intended range. Third party additions should pass the same testing scrutiny as the rebreather itself. Modifications to your rebreather may void your warranty. Some Authorized Service Centers are unable to support modified rebreathers, so think hard before making any changes to the tested and validated design of your equipment.

An inline slider valve cuts off OC gas to the BOV, and may cause a need for further modifications such as an OPV on the first stage.

Specialty Equipment

Specialty CCR diving generally requires a little more equipment and may include items such as:

✓ Low profile mask and fins (straps taped or spring straps)

✓ Optional back-up mask

✓ Type T rebreather

✓ Suitable exposure suit for deeper, longer dives

✓ Bail-out cylinder(s) with a first and second-stage, submersible pressure gauge, LP hose for dry suit and/or offboard gas connection. Long hose is optional, but minimum 5-foot length is recommended and up to 7-foot may be required. I recommend using two bailout tanks for balance and to offer redundancy. These tanks will be matched to penetration goals and diver surface air consumption rate (SAC rate) on open circuit. Some divers use 2 x 40 cft and others who wish to swim further into the cave or deeper depths select 2 x 80 cft. Additional bailout tanks and mixes may be required.

✓ Suitable wing and harness for rebreather ("back-mounted wing" is recommended). Redundant buoyancy may be provided by a dual wing or drysuit.

CHAPTER 4 - TRAINING

- ✓ Primary light with appropriate intensity and burn time for the dives planned (for cave, wreck and some deep applications).

- ✓ Two battery-powered backup diving lights

- ✓ One primary line reel with minimum 350 feet/ 110 m of guideline per team.

- ✓ Safety reel with at least 100 feet/30 m of guideline. Optional back-up safety reel.

- ✓ DSMB with reel or spool for open water. Optional backup.

- ✓ Backup dive computer or timing device (CCR compatible), depth gauge, slate, pencil and submersible dive tables. Redundant dive planning capability.

- ✓ A small knife or a "Z" knife line cutter with back-up line cutter or shears.

- ✓ At least three directional line markers such as "arrows." At least two non-directional line markers such as "cookies" or clothes pins (for cave classes).

- ✓ In-water decompression cylinder for dives in which decompression may be a factor. The cylinder will incorporate all support equipment including, but not limited to, regulators(s), submersible pressure gauge, adequate gas for at least one and a half times the gas volume required for the expected needs of the dive team. Optional off-board LP hose if applicable.

- ✓ Jump/gap reel with minimum of 50 feet/ 16 m guideline (for cave classes)

- ✓ Ancillary gear including tools, checklists, spare parts, laptop, calculator, etc.

- ✓ John line (for open water)

- ✓ Oxygen Analyzer

- ✓ Emergency oxygen kit (may be provided by instructor)

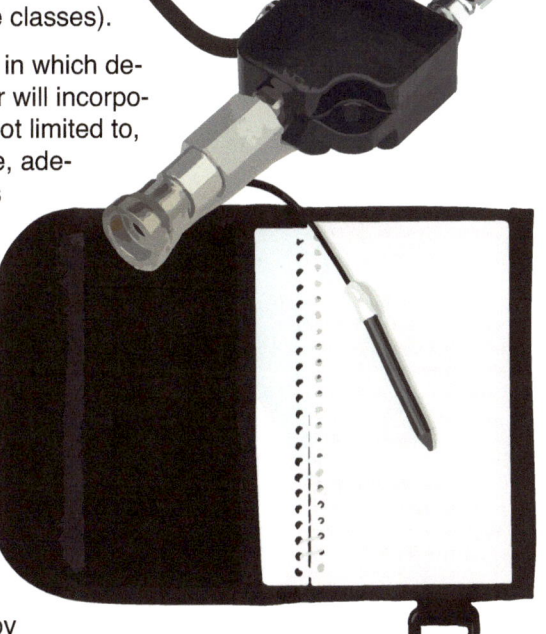

[4] PADI Tec 40 CCR Diver Manual, Chapter One, Basic Functions of CCRs and SCRs

5

Physics

In this chapter:
- *Calculating Pressure*
- *Using Dalton's Law*
- *Rebreather Setpoint*
- *Selecting Bailout Gas*
- *Equivalent Air Depth*

Physics for Rebreather Divers

There is a little bit of math involved in rebreather diving. I've never been much of a numbers person, practically failing my high school Calculus class. However, it's important to understand some basic equations and know where to find them if you need them. Trying to commit formulae to memory is not always a great idea. Sometimes it is better to look things up. Keep your references handy.

Calculating Pressure

The partial pressure of a gas in a breathing mixture is measured in atmospheres absolute or bar (ATA/bar). As a qualified diver, you already know that the deeper you go, the more pressure there is. Dalton's Law (also called Dalton's Law of Partial Pressures) states that the total pressure exerted by a mixture of gases is the sum of the pressures that would be exerted by each gas if it alone were present and occupied the entire space. In rebreather diving we refer to partial pressure of oxygen (PO_2) frequently. Dalton's Law can be expressed as:

P(total) = P1+P2+P3...Pi

Generally speaking, in air at sea level:

PO_2 = 0.21 ATA/bar

PN_2 = 0.79 ATA/bar

Total Pressure = 0.21 + 0.79 = 1.0 ATA/bar

CHAPTER 5 - PHYSICS

I've always remembered Dalton's Law by visualizing a rogue band of cowboys, "Dalton's Gang." The gang is made up of a bunch of individuals who are all present in the gang, exerting their own pressures and opinions on the group. It's a simple visualization, but for me, much easier than recalling figures.

Partial Pressure of Oxygen at Various Depths

Depth fsw	Depth msw	Pressure ATA/bar	PO_2 air diluent	% O_2 air diluent
0	0	1	0.21	0.21
33	10	2	0.42	0.21
66	20	3	0.63	0.21
99	30	4	0.84	0.21
132	40	5	1.05	0.21
165	50	6	1.26	0.21
198	60	7	1.47	0.21
300	90	10	2.1	0.21

Using Dalton's Law

In rebreather diving, there are many ways that we use Dalton's Law.

Converting Depth into a Pressure Value

To convert any depth into pressure, use these simple formulae:

$$P(bar) = \left[\frac{Depth\ (msw)}{10}\right] + 1$$

$$P(bar) = \left[\frac{Depth\ (fsw)}{33}\right] + 1$$

CHAPTER 5 - PHYSICS

Choosing the Maximum Operating Depth (MOD) of Bailout or Diluent

Using 1.4 for the Target Max PO₂, you can calculate your MOD conservatively with this formula:

$$P(bar) = \frac{Target\ Max\ PO_2\ (P)}{Fraction\ of\ O_2}$$

Then to convert the pressure to an actual depth:

METRIC:

IMPERIAL:

$$\left[P(bar) \times 10\right] - 10 \qquad \left[P(bar) \times 33\right] - 33$$

Choosing the Ideal Bailout Gas

The ideal bailout gas needs to be breathable at your maximum depth. If the maximum depth is only momentary, you may choose to use 1.6 in your calculations. If you are like me and dive in a cave, or want some more conservatism built in, then you should use 1.4 for the max PO₂ of your bailout gas.

The higher the PO₂ you choose, the faster you will clear any decompression obligations. This might be important in the emergency scenario of an open circuit abort, as will the gas supply savings based on faster decompression. Use the following formula to make that calculation.

$$Fraction\ of\ Oxygen\ (FO_2) = \frac{Target\ Max\ PO_2\ (P)}{Depth\ (P)}$$

Rebreather Setpoint

In rebreather diving we use a term called "setpoint." This refers to the desired PO₂ in the breathing loop. In an eCCR, PO₂ setpoint is either auto selected or chosen by the diver and is used to determine when the solenoid valve injects oxygen into the breathing loop. Oxygen sensors detect a PO₂ drop and subsequently trigger an electromechanical solenoid valve to inject oxygen. In an mCCR, divers also use the term setpoint when referring to the level of oxygen they target with their manual injections.

CHAPTER 5 - PHYSICS

Percentage of Oxygen in a Breathing Gas at a 1.2 Setpoint

As mentioned earlier, CCRs are like carrying a gas mixing station on your back. Looking at the chart below, you can see how a PO_2 setpoint of 1.2 would offer a percentage equivalent of a tremendously rich nitrox in shallow depths.

Depth fsw	Depth msw	Pressure ATA/bar	PO_2 bar	FO_2 %
10	3	1.3 bar	1.2	92%
20	6	1.6 bar	1.2	75%
30	10	2.0 bar	1.2	60%
60	20	3.0 bar	1.2	40%
100	30	4.0 bar	1.2	30%
130	40	5.0 bar	1.2	24%
165	50	6.0 bar	1.2	20%

Floating Setpoint

In the chart above, you can see that it would be wasteful to use a setpoint of 1.2 in 10 fsw. You don't really need to be breathing the equivalent of EAN92 on a shallow, no stop dive! Furthermore, if your rebreather was set to 1.2 on the surface, the electronics package might continuously inject oxygen, since it could never reach a theoretical 1.2 setpoint. For this reason and others, some rebreathers offer a feature called a "floating setpoint." This is sometimes called auto setpoint, ASP or dynamic setpoint.

A floating setpoint is a setpoint that uses logic to efficiently and automatically switch through reasonable set-

On a manual rebreather, you may be monitoring your display and maintaining a floating setpoint since you may need a lower setpoint in shallow water in order to utilize your gas supply judiciously.

points throughout a dive. eSCRs such as the Hollis Explorer use a floating setpoint to find an optimized balance between lengthening your dive and reducing the likelihood of decompression all while considering exposure to oxygen and remaining gas supply. Manual CCR divers work at a lower setpoint on the surface and gradually increase their oxygen levels to a higher setpoint on the bottom. They are manually controlling a floating setpoint.

On many rebreathers you will need to manually select and confirm your high setpoint at depth.

Some rebreathers use Automatic Setpoint (ASP) in a slightly different way. The diver chooses a shallow setpoint of perhaps, 0.5. They also choose a maximum setpoint of perhaps 1.2. When the ASP function is selected, their rebreather will slowly ratchet up a floating setpoint until they reach their maximum setpoint value. This is designed so that a diver who is descending on a low setpoint can allow themselves to get distracted while they work on issues such as clearing their ears. The rebreather will not allow the actual PO_2 to drop back to the low setpoint, but continues to hold or climb until they reach maximum depth where it chooses the high setpoint. Still other rebreathers will manage an automatic floating setpoint on ascent as well, handled either completely automatically, or by depth trigger points that change the setpoint as a diver reaches that waypoint.

Choosing a Setpoint

In your recreational nitrox training, you should have learned that the accepted maximum PO_2 for a breathing gas is 1.4 ATA/bar with an absolute max operating depth at 1.6. If you are technically oriented, then you might have used a decompression gas with a PO_2 max of 1.6 such as using oxygen at 20 feet/6m. In CCR diving we operate on lower setpoints, generally with a maximum of 1.1 to 1.3. This lower PO_2 offers you a better safety envelope since there can be swings in PO_2 as the oxygen is dynamically added and mixed in the loop. Your body is also subjected to greater oxygen exposures on a CCR dive and you need a PO_2 setpoint that will help you remain inside safe oxygen exposure limits

CHAPTER 5 - PHYSICS

(more on this in next chapter). Furthermore, oxygen sensors have some variability, especially if they are moist. A wet sensor face tends to cause the sensor to read a little low and a little slow, therefore your actual PO_2 may be higher than you think. Rapid descents can also cause the PO_2 to rise quickly or even spike above the setpoint. This is another reason why a setpoint in the 1.1 to 1.3 range is preferred.

Surface Setpoint

You will use a low setpoint on the surface, somewhere in the range of 0.5 to 0.7. The maximum achievable setpoint on the surface is 1.0, but that can only be reached if the entire loop is voided of other gases and pure oxygen exists alone in that space. As you can imagine, that is a good reason not to use an unnecessarily high setpoint on the surface. Trying to reach it would be very wasteful.

The safe range of PO_2 for the human body is generally considered to be between 0.16 and 1.6. We choose a setpoint of 0.5 to 0.7 for the surface in order to account for task loading and distraction. A PO_2 of 0.5 to 0.7 allows plenty of time to detect problems (such as failing to turn on the oxygen tank) before PO_2 can drop to level that could induce hypoxia.

We don't want the setpoint to be too high, since as you descend, the PO_2 in the loop will rise from your descent through increased pressure gradients. If your descent is slow to normal, the rebreather injections may match your metabolism of oxygen. But if your descents are a little faster, you'll see the PO_2 rise on the way down. In a CCR, the ideal situation is created when the low setpoint is left as a safety net on the way down and the actual PO_2 slowly rises to the planned maximum

Depending on the model of eCCR, you may have to switch your setpoint to the low setpoint when you reach the surface. If you fail to lower the setpoint, the solenoid might continue to fire and add oxygen to achieve a high setpoint that simply can't be reached. The Hollis Explorer eSCR (below) automatically floats the ideal setpoint and adjusts for an appropriate surface setpoint.

CHAPTER 5 - PHYSICS

setpoint at the bottom. On the bottom, the diver either manually moves to the high setpoint or the unit electronically switches to the high setpoint.

Your eCCR backup computer, or primary computer on an mCCR, won't likely have a floating setpoint option. Some will provide a trigger depth where the computer is instructed to switch from a preset low to a preset high setpoint. You can also leave the computer on the low setpoint for descent, switch to high on the way down when you reach that value and then switch back on ascent where appropriate. This gives you a conservative dive profile. For SCR divers, the backup computer can be a simple nitrox computer that is set for the target fraction of oxygen at depth, however you won't be gaining any no deco time on your SCR with this method.

Equivalent Air Depth (EAD)

Equivalent Air Depth is the depth at which a diver breathing air would be taking on nitrogen at an equivalent rate while breathing nitrox at some other depth. This was a critical concept in early nitrox diving since all divers used tables for dive planning and nitrox tables were non-existent, let alone decompression diving software. Although the concept is still important to understand, you will rarely need to use this math, unless you are using an air diving computer to try to conduct an SCR dive.

METRIC:
$$EAD = \left[\frac{(1-FO_2)(D+10)}{0.79}\right] - 10$$

IMPERIAL:
$$EAD = \left[\frac{(1-FO_2)(D+33)}{0.79}\right] - 33$$

In a rebreather loop, your EAD should be less than your actual depth. The EAD changes through the dive as the depth or setpoint changes. In order to calculate EAD for a given depth, you would need to use the fraction of oxygen that is in your loop as opposed to what is contained in the diluent bottle.

Physiology

In this chapter:

- *Respiration*
- *Work of Breathing*
- *Oxygen Metabolism*
- *Problems*

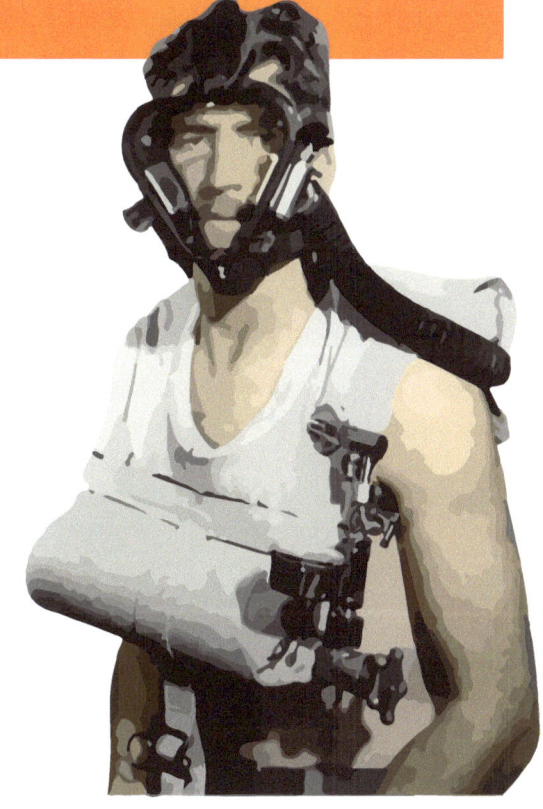

Respiration

Your first breaths on a rebreather will feel quite unusual. With open circuit SCUBA, you are accustomed to a certain amount of effort being required to open the demand valve of the regulator. Breathing on CCR feels a little more natural. You'll have to stop habits that you might have gathered in open circuit such as holding your breath or inhaling to rise through the water column. Breathing resistance will change, depending on where the counterlungs are mounted on your body and how your body is oriented in the water column. If the counterlungs are over-the-shoulder style (OTS), they may breathe similarly in all body positions except directly head-down, where they may get squeezed. Back-mounted counterlungs will be slightly harder on inhalation when you are horizontal and front-mounted counterlungs will breathe easier on inhalation. Envision where your body and the counterlungs are in relation to the water column. Are you exhaling into higher pressure or lower pressure? You may notice changes in breathing resistance and buoyancy when you roll over or move from horizontal to vertical for ascent. More on that later.

Water in your breathing loop affects breathing resistance. Even if your loop is secure and mouthpiece habits are good, you will be exhaling moisture into the loop. Since it is trapped in the closed loop, it accumulates. If there is a lot of moisture running around in your exhalation hose you will likely hear gurgling. Accumulated metabolic moisture can make it difficult to breathe if the "lung juice" partially blocks the non-return valve. You will learn how to clear the loop in class.

CHAPTER 6 - PHYSIOLOGY

As you dive the rebreather, you will notice the difference in humidity as compared to open circuit diving. You won't feel that "cotton-mouth" sensation you might be accustomed to. You may also notice that the gas feels warmer. The exothermic reaction in the canister produces moisture and heat. To me, that makes for a more comfortable dive and may reduce the risk of DCI, however those issues are in need of scientific study.

Work of Breathing

Work of breathing (WOB) is a specific term to describe the "feeling" or effort for inhalation and exhalation through an entire cycle of breathing, but there is a lot more to it. Work of breathing is measurable and can be charted on a graph. The results of those tests should be made available by the manufacturer. These tests are required for CE (European Commission) certifications, which are in turn required for rebreathers sold in the European Union (EU). More on testing protocols later in Chapter 10.

One aspect of work of breathing is called "resistive work of breathing." This describes how the gas flows through the unit and takes into account things like the size of hoses and orifices and everything else that might generate a resistance to breathing. Such things as a small mouthpiece, poorly designed mushroom valves, small hoses or counterlungs and gas density can degrade the resistive work of breathing. Other factors will affect work of breathing, such as depth (deeper dives net denser gas therefore net higher resistive WOB) or workload.

It's like announcing to a crowd in a theater that there is an emergency. If you calmly ask people to leave the room in an orderly fashion, they'll likely get out the exits efficiently. If you

Over-the-shoulder (OTS) counterlungs (above) and back-mounted counterlungs (below).

scream and panic, the crowd may run into chairs and seats and get jammed up in a narrow doorway. If you add twice the people to the crowd, like twice the molecules of denser gas at depth, then they have an even harder time getting out the door. If the exit door is jammed slightly, like a poorly designed mushroom valve in a rebreather, then it is even tougher to escape.

Resistive WOB can affect the short-term "feeling" on a rig - breath by breath - but may also generate a different result over the timeline of a dive profile. If you have ever used a counterlung that was too small, you know the feeling… air starvation.

If your rebreather breathes great in your garage, it may not feel the same underwater. This is where the second half of the work of breathing test comes in to play: hydrostatic work of breathing.

CE standards also test for hydrostatic work of breathing. These tests

Work of breathing (WOB) is the effort required to complete a cycle of breathing. Work of breathing can be affected by breathing hose diameters, check valve design, scrubber design, counterlung placement and design, depth, absorbent material and other factors. Photo: Mark Long.

quantify how a rebreather breathes in the water column under increased pressure in different positions. It might breathe great while riding an exercise bicycle on the surface, but once fitted to the diver, the position of the equipment has to work too. The diver's orientation in the water column can affect this test, so several pre-determined, calibrated positions are tested.

The type of counterlungs, their position on the diver's body and personal fit of gear can affect the hydrostatic WOB. Let's look at different types of counterlungs and the forces we are fighting to overcome.

Chest mounted: Easy inhalation, harder exhalation (face down position)

These counterlungs, such as on the Lar-5 military CCR, are mounted on the chest. In a horizontal position, they are under greater pressure than the diver's lungs as measured from the centroid (the center point of the lungs for measuring purposes).

Back mounted (Rear mounted): Harder inhalation, easy exhalation

CHAPTER 6 - PHYSIOLOGY

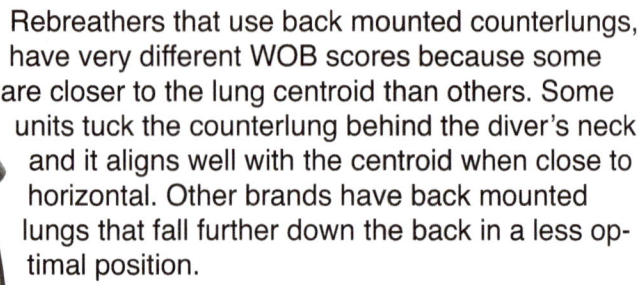

Rebreathers that use back mounted counterlungs, have very different WOB scores because some are closer to the lung centroid than others. Some units tuck the counterlung behind the diver's neck and it aligns well with the centroid when close to horizontal. Other brands have back mounted lungs that fall further down the back in a less optimal position.

Over the shoulder (OTS): Easy hydrostatic work of breathing

These counterlungs should be ideal and easiest to breathe, but the resistive work of breathing often negates the gains of good positioning close to the lung centroid. They tend to score poorly in the resistive WOB since the gas travels through a lot of turns and bends. It's all in the complexities of the design.

CE testing takes into account a combination of resistive and hydrostatic work of breathing in results. A great score on one side and a poor score on the other may offer an overall average score. Results are additive, so it is important to review the results of both tests to fully quantify work of breathing of your rig.

Oxygen Metabolism

During your breathing cycle on the surface, you inspire about 21% oxygen. About 4% passes into the body to fuel oxidative metabolism. Roughly 17% oxygen is contained within an exhaled breath at sea level. One of the functions of your rebreather is to replace the lost oxygen either manually or automatically. The term Respiratory Minute Volume (RMV) refers to the total volume of gas inhaled and exhaled during one minute of normal breathing. The RMV is determined by respiratory rate and volume of gas ventilated in each breath. Some people breathe more rapidly and others have larger lungs or more cells to feed. Still others have a high metabolic rate. In open circuit diving, these people will use more gas and their RMV often dictates when the dive is ended. It means that your buddy who is the size of an NFL linebacker may use up his open circuit gas faster than you.

In most SCR diving, the dive time is controlled by the flow rate on the rebreather. Some SCRs are very efficient, such as the Hollis Explorer, but other styles, such as the Dräger Ray operate on a continuous flow that uses gas faster. If both you and your linebacker friend are diving SCRs, you may find that your dive durations are similar.

CHAPTER 6 - PHYSIOLOGY

The available dive time on a CCR can be determined by metabolic rate or workload. The harder you work underwater, the faster you will metabolize oxygen molecules and need them replaced. A relaxed dive versus an active dive in current will change the overall range of the oxygen tank. However, no matter how deep you dive, the oxygen tank will generally last the same amount of time. That's right! Unless you are being wasteful or have a leak, the tank will last the same time in 10 feet of water as it will on a deep dive! Other consumables may control the final dive time if they are expended faster than oxygen supply.

Average Oxygen Consumption Chart

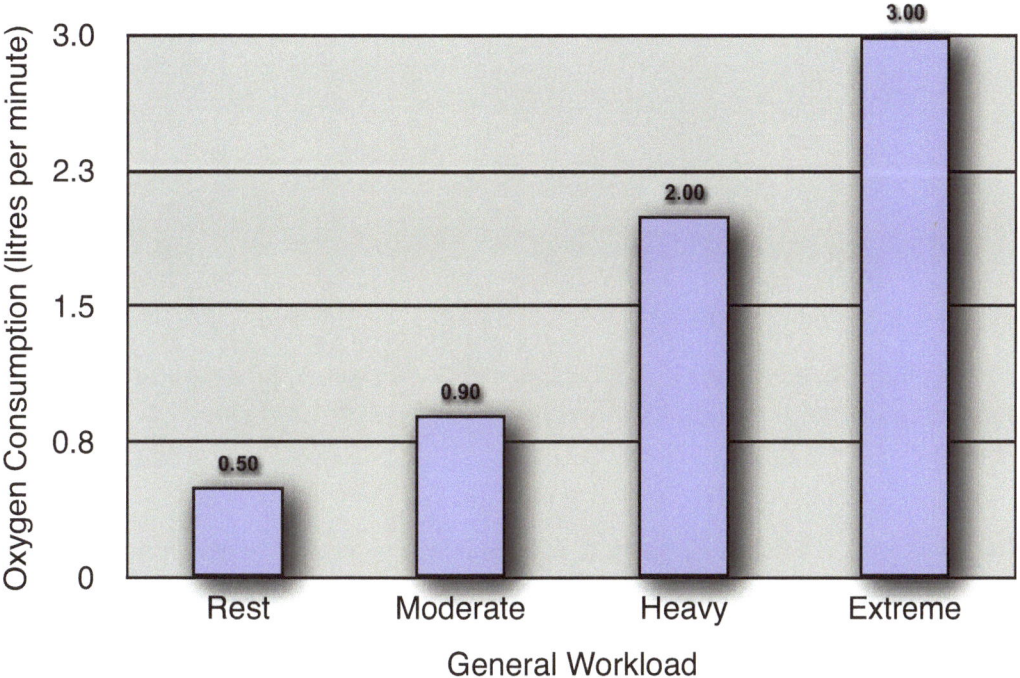

Oxygen consumption on a dive may be affected by other factors. Anything that causes the diver to vent gas will raise the overall consumption rate. If your dive profile goes up and down a lot, then you will be dumping gas as you ascend and replacing it as you descend. This will net an overall higher gas consumption in both diluent and oxygen tanks. Leakage through an ill-fitting mask or frequent mask clearing will also cause you to vent bubbles and require gas replacement. Therefore, if we assume that there is no leakage from the mask or loop, then the oxygen tank duration is truly independent of depth. The chart on the following page makes a comparison of open circuit gas usage to closed circuit oxygen usage.

CHAPTER 6 - PHYSIOLOGY

Comparison of Open Circuit Gas and Closed Circuit Oxygen Use

Depth fsw	Depth msw	Pressure ATA/bar	Open Circuit l/min.	Closed Circuit l/min.
0	0	1	25	1
33	10	2	50	1
66	20	3	75	1
99	30	4	100	1
132	40	5	125	1
165	50	6	150	1
198	60	7	175	1
300	90	10	250	1

Let's look at just how long your onboard tank can last compared to a standard open circuit SCUBA cylinder. Many rebreathers operate on a small oxygen tank filled to 200 bar (3000 psi) pressure.

O_2 supply = 2 liters @ 200 bar = 400 free liters available

Metabolic rate = 1.0 lpm @ oxygen metabolism at a moderate workload

Therefore the O_2 supply duration in this tank can last up to 400 minutes irrespective of depth.

In open circuit diving, if we assume a low effort and usage of 20 lpm of open circuit gas RMV at the surface, then...

Open circuit supply @ surface = 20 minutes

Open circuit supply @ 20 m = 6.66 minutes

Therefore, in this case:

6.66 mins open circuit = potential of 6.66 hrs on a CCR!

Problems

In open circuit SCUBA diving, (with the obvious exception of breathing the wrong gas) if you can breathe in and out, everything is fine (most of the time). With a rebreather, this is not necessarily the case. You may have adequate volume to breathe in

CHAPTER 6 - PHYSIOLOGY

and out of your loop comfortably, but the gas can be toxic and lethal. Gas supply emergencies in open circuit are usually accompanied with a "boom-hiss" which scares the diver into action. This is not necessarily the case in rebreather diving. An oxygen tank can be quietly depleted and depending on the brand of rebreather, there may be no warning at all.

Hypoxia (O_2 deficiency)

Hypoxia is defined as lack of oxygen at cellular level. The first signs may occur at a PO_2 of 0.16 ATA with serious signs when it drops to PO_2 of 0.10 ATA or lower. Every individual is a little different and additional factors may contribute to a higher or lower threshold for consciousness on any given day. Symptoms of hypoxia include: lack of coordination, euphoria, inability to think clearly, unconsciousness and death, without any ability to react. The cellular oxygen level does not necessary affect the brain like an on and off switch. As cellular oxygen drops, the level of consciousness may ebb too until complete unconsciousness results. For this reason, you can't count on being able to detect signs and react to developing hypoxia. Hypoxia is a major concern. Because it can creep up without a diver's knowledge, closely monitoring displays is the most important prevention.

It may be hard to believe, but many, if not most, hypoxia incidents have occurred because a diver failed to activate their rebreather in some way. One very common failure is to neglect turning on the onboard tank valves. There are several actions/inactions that can lead to hypoxia:

* Failure to turn on oxygen tank
* Depleted onboard oxygen
* Stuck diluent add valve
* Failed diluent first stage
* Electronics failure
* Controller not activated
* Rebreather turned off
* Oxygen supply selector turned off (in some units)
* Failure to analyze onboard gases
* Using the wrong gas
* Calibration issues
* Diluent leaking into loop

Technical Instructor and CCR pioneer Phil Short demonstrates the greatest single preventive measure for rebreather issues ranging from hypercapnea to hypoxia: complete a pre-dive check and pre-breathe with your nose blocked.

CHAPTER 6 - PHYSIOLOGY

* Bad sensors
* Solenoid valve failed in closed position due to contamination in hoses
* Rapid ascent
* Dead batteries in rebreather
* Loss of controller
* Failure to monitor displays
* Failure to add oxygen (manual ops)
* Blocked orifice (Active SCR)
* Ignoring alarms (some CCRs have low gas, low PO_2 and other alarms)
* Breathing the loop on surface when rebreather is off

Consistent use of printed or automated checklists are key to safe operations.

Notably, almost all of these issues can be prevented by completing a thorough pre-dive check and pre-breathe. The above list represents some of the most common mistakes. These bad habits are correctable. Accidents are preventable with good, consistent safe diving procedures.

Hyperoxia (CNS O_2 toxicity)

Hyperoxia is defined as an excess of oxygen at the cellular level. Central nervous system toxicity (CNS toxicity) symptoms generally result from PO_2 levels of greater 1.6 ATA/bar. In some cases toxicity may occur when PO_2 is greater than 1.0 ATA/bar.

There are several contributing factors which can affect the likelihood of having a problem with CNS Oxygen Toxicity:

- Actual PO_2 breathed (affected by setpoint or manual additions)
- Duration of exposure
- Level of exertion
- Cumulative O_2 exposure
- Other less studied factors such as carbon dioxide retention may also reduce seizure threshold.

There are numerous symptoms

Even though you will use onboard supplies slowly, you still need readable SPGs for both the oxygen and diluent tanks. You must be able to monitor these during your dive.

CHAPTER 6 - PHYSIOLOGY

which may, but not necessarily, precede unconsciousness that leads to a high likelihood of drowning and death:

- CON - Convulsions
- V   - Vision (abnormalities such as seeing spots)
- E   - Ears (hearing disturbances such as ringing)
- N   - Nausea
- T   - Twitching
- I   - Irritability
- D   - Dizziness

CNS tracking couldn't be simpler. I'll cover this in more detail on page 124 in the book, but in general, if you use the NOAA O_2 exposure table, at 1.2 ATA you have 3.5 hours dive time or 4 hours per day available to keep you within safe CNS limits. However, if you push the PO_2 up to 1.5 ATA, the dive limit is reduced to 2 hours!

There are several reasons why a hyperoxic loop may develop:

✻ Rapid descent
✻ Solenoid valve failed in open position
✻ Manual oxygen valve stuck open
✻ First stage high pressure seat failure in oxygen tank resulting in free flow
✻ Leaking oxygen gas block
✻ Injecting too much oxygen
✻ Pressing the wrong injector button
✻ Hooking up wrong supply hose to ADV
✻ Faulty sensors
✻ Old sensors
✻ Current limited sensors causing a "false-low" reading
✻ Wet sensors
✻ Bad calibration
✻ Wrong mix in SCR
✻ Too deep for diluent supply or SCR gas
✻ Gas not analyzed
✻ Diluent tank filled with oxygen
✻ DIluent tank swapped with oxygen tank

CCR divers begin their dive with a low setpoint because the PO_2 in the loop will rise as they descend. A very rapid descent may require the diver to flush with diluent or pause at an intermediate depth to keep the PO_2 within its safe upper limits.

CHAPTER 6 - PHYSIOLOGY

Whole Body Toxicity

Whole body oxygen toxicity (also referred to as pulmonary toxicity) is different from CNS toxicity. You may have learned about this in a nitrox class and assumed that it would never affect your dive profiles. In the past, it was primarily commercial and expeditionary divers that had to consider whole body toxicity. This type of toxicity is possible at low PO_2 levels of 0.5 to 1.0 ATA/bar or higher over very long durations. These types of exposures can occur on long diving trips and expeditions where your daily oxygen exposures add up over time.

Also known as the Lorraine-Smith effect, pulmonary toxicity symptoms may begin as a lung irritation or a mild tickle and progress to sub-sternal pain, burning and coughing. Some divers have reported other effects from extreme, long expeditionary dives or chamber treatments. These symptoms may eventually lead to inflammation and edema with uncontrollable coughing and shortness of breath.

With dive missions over 21 hours in duration every four days, several exploration divers on the Wakulla2 Project experienced Pulmonary Toxicity symptoms. Photos: U.S. Deep Caving Team, Inc.

At Wakulla Springs we used a pressurized capsule that could be hoisted to the surface and mated with a recompression chamber to keep us pressurized for lengthy decompression times. It gave the team peace of mind for conducting unconventional oxygen exposures.

When I participated in the United States Deep Caving Team's Wakulla2 Project, expedition divers were exposed to high partial pressures between 1.0 and 1.6 for over 20 hours per mission, repeating those missions every fourth day. I experienced the irritating cough of pulmonary toxicity while some of my colleagues reported hot red finger tips and toes in addition to burning chest irritation. Symptoms subsided over time after the project was completed with nobody reporting any lingering issues. Despite the fact that the symptoms abated, it is important to note that any efforts at recompression therapy during that time could have been compromised. Divers

CHAPTER 6 - PHYSIOLOGY

with DCI are treated with hyperbaric PO_2 exposures. If you are already experiencing pulmonary toxicity symptoms, recompression therapy may result in further harm. Fortunately, it is possible to track pulmonary exposure and "leave room" for any necessary recompression therapy within your dive plan. This chart is included in the procedures section of the book.

There are several situations that could lead to development of pulmonary oxygen toxicity:

* Setpoint too high for dive durations
* Long dive trips
* Multi-day, multi-dive exposures
* Unexpectedly long decompression obligation

Inside the Ice

Immersed in the hostile yet stunning world that is Antarctica, one can only survive under totally artificial means. In every aspect of traveling to the southern continent, we needed technology to ensure a synthetic environment to keep us warm and safe. In 2000, Wes Skiles, Paul Heinerth and I traveled to Antarctica with a goal to be the first people to cave dive inside an iceberg.

We had applied for an expedition permit from the National Science Foundation (NSF) to use rebreathers for our work. After failing to convince the NSF of the safety benefits of rebreathers, we were fortunate to get a permit from authorities in New Zealand. The only member of the team with a background in ice diving, I insisted we use rebreathers on our project. I knew they would keep us warm and should prevent free flow issues in the water that was a mere -1.8°C (28° F).

On the first exploration of an iceberg cave, Paul and I cautiously entered a deep underwater crevasse and found a gaping fissure that extended out of sight. Sheer white walls dropped interminably in a narrow crack. We swam into the fracture and drifted down to the sea floor. As we hit 40m/130 feet, we discovered that the berg was undercut and we could continue our swim below the mass. We found a dazzling world of colorful tunicates, sea stars and curious creatures. Brilliant reds cast a glow on the underside of the ice just a few feet over our heads. The forces of strong currents carved conduits and passageways through the ice and brought nourishment to the plentiful life. Large scalloped hollows textured the walls

CHAPTER 6 - PHYSIOLOGY

like dimples on a giant golf ball. It was similar to the karst environments that we were familiar with, only it was made of ice.

We slipped silently through the underbelly of the iceberg with our CCRs, hearing only the occasional fire of the solenoid valve. We had penetrated a full quarter of the berg when we decided to turn around. I heard a faint moaning reverberate around me. Not recognizing the sound, I carefully inspected my gear. Finding nothing awry, we turned to swim back to the waiting Zodiac.

As Paul and I hovered at our decompression stop, I noticed the terrain had changed. The entrance looked significantly different than when we began our dive. As we snaked our way out through the ice, I looked up to our waiting boat and saw Wes Skiles and First Mate, Matt Jolley in the midst of obvious celebration. I later learned that they had been frantic. During our dive, a deep and frightening groan issued forth from the iceberg. A large piece of ice in the opening calved and rolled sending them on an eight-foot swell up, and a sixteen-foot headlong crash back down again in the Zodiac. They were happy to see us alive.

Looking back on that and other experiences in Antarctica, I am certain it was only because of the rebreathers that we came home alive from that trip. They allowed us time to solve problems. They kept us warm through long dives, allowed for high workload diving and were resilient to free flow gas emergencies.

Hyperoxic Myopia (Lenticular Oxygen Toxicity)

Hyperoxic myopia, or nearsightedness, has occurred in rebreather divers after prolonged exposures.[5] It has also been documented in patients undergoing lengthy hyperbaric treatments. It is normally reversible. Filmmakers Howard Hall and Bob Cranston recounted having this problem after a lengthy and deep IMAX film shoot using rebreathers. They described going to the airport and having difficulty reading the signs directing them to their departure gate. Both divers described a full recovery after a period roughly equivalent to the underwater shoot itself.

Middle Ear Oxygen Absorption Syndrome

Hours after completing a dive, some rebreather divers have reported ear squeeze injuries. Upon surfacing, your ear's eustachian tubes may be filled with a relatively high FO_2. Oxygen molecules in the middle ear will slowly be absorbed into tissue, but are not always replaced as fast with nitrogen. Pressure outside the eardrum becomes greater than inside and a squeeze results. If you

CHAPTER 6 - PHYSIOLOGY

feel the urge to gently equalize after diving, you will clear up this problem. If you are unable to equalize, a squeeze barotrauma can result.

Hypercapnia (CO2 Toxicity)

Hypercapnia (also known as hypercarbia) is defined as excess CO_2 at the cellular level. CO_2 is generated by oxidative metabolism and is normally expelled through the lungs into the breathing loop. In the scrubber, it binds with the absorbent material if everything is working properly. Hypercapnia triggers a reflex which increases the diver's breathing rate and access to oxygen. As the CO_2 level rises, several symptoms may occur:

0.02 ATA/bar - doubles breathing rate (dyspnea)

0.06 ATA/bar - distress, confusion, lack of coordination

0.10 ATA/bar - severe mental impairment

0.12 ATA/bar - loss of consciousness, death

Divers have additionally reported muscle twitching, facial tingling, flushing, headache, lethargy and tunnel vision.

Some rebreathers are equipped with CO_2 alarms that notify the diver if their scrubber has failed. The alarms are often set at 5 and 10 mb thresholds for warning and bailout respectively. These alarms are a good safeguard to warn a diver of an improperly packed scrubber or similar issue, but do not address issues of diver CO_2 retention, which can only be detected by reading end tidal CO_2 levels in the mouthpiece. This would effectively measure the very end of an exhaled breath before it had a chance to mix/dilute with other gas. This technology is not currently available.

My diving partner Brian Kakuk hovers on Challenger Sea Mount at almost 450 feet of depth. We had an aggressive and physical dive plan but had to be careful to minimize any overexertion that could lead to carbon dioxide issues. Swinging a sledgehammer, Brian had to bring up rock samples and biological specimens for scientists on this NOAA expedition.

CHAPTER 6 - PHYSIOLOGY

The following situations may lead to hypercapnia:

* Failure to load a scrubber into the rebreather
* Loading an empty scrubber canister
* Reusing old absorbent material
* Improper packing of a scrubber
* Repacking scrubber with expired media
* Mixing fresh with partially used media
* Channeling of scrubber material
* Damage to an ExtendAir canister (if applicable rebreather)
* Failure of the O-ring or seal on the scrubber canister
* Improper assembly of rebreather
* Overexertion resulting in scrubber by-pass
* Exceeding depth limits of a scrubber
* Using a scrubber beyond its recommended duration
* Failure to account for shorter duration of scrubber lifespan due to depth (scrubber roll off)
* Improper scrubber media for deep dive
* Using rebreather beyond tested limits
* Wet scrubber material
* Flooded canister
* High work of breathing
* Poorly designed or untested BOV
* CO_2 buildup in a full face mask
* Untested modifications to breathing loop
* Poorly serviced open circuit regulator or BOV
* Breath holding (skip breathing)

High workload dives are more likely to create scenarios where divers can over breathe their canister's scrubbing capability. Scrubber duration is lessened during deeper dives. Ask the manufacturer of your rebreather about the specifications for diving at 100m depth.

CHAPTER 6 - PHYSIOLOGY

The problem with hypercapnia is that it can easily be overlooked when a diver is swimming at a high workload. Scrubbers have their limits, and if you push too much CO_2 through them, they may not be able to do their job. This is referred to as "bypass." If carbon dioxide is allowed to bypass, then it will build up in the diver's body. As this occurs, the diver can experience a shortness of breath that is easily overlooked when working hard. It creates a vicious cycle that can be unrecoverable. If the diver experiences shortness of breath, he may choose to bail from the rebreather. With the CO_2 still high in the body, the diver might not sense any relief from his open circuit regulator. Several accidents and fatalities have occurred where it was noted that the diver repeatedly switched from closed circuit to open circuit, unable to catch his breath. By the time he reacted to symptoms, it was already too late. The lesson is not only prevention of scrubber problems, but also highlights the importance of bailing out at the earliest signs of respiratory distress or other symptoms. A carbon dioxide breakthrough is considered a non-recoverable loop failure. You will not solve the problem through lower workload or lower respiration rates.

Proper packing of a rebreather canister takes time and patience with an attention to cleanliness. When you keep your work area tidy, you will avoid grinding granules of sorb into the bottom of the canister body.

Scrubber duration can be predicted in several ways: duration testing by the manufacturer, tracking the thermal imprint of the reaction zone, tracking the amount of oxygen consumed by the diver and CO_2 monitoring in the loop.

In early rebreather training classes, divers were challenged to dive a CCR in a pool without a scrubber installed. I recall working with Tom Mount at his home. The students were aware that there was no scrubber in their rebreather. They were instructed to swim hard with their hands against the wall of the pool. Tom and I stood on either side of the diver while he was instructed to bailout at the first sign of a problem. They were timed to see how long they could swim before they either bailed to open circuit or passed out. It was a bit of a failed experiment since the entirely male class seemed to be competing for honors as the diver most able to cope with high CO_2! One by one the divers took their

CHAPTER 6 - PHYSIOLOGY

turns. Not one actually bailed to open circuit, though several later reported facial twitching, difficulty getting a satisfying breath and warm flushing of the face. Most started to display strangely grotesque leg kicking at around 45 seconds, at which point we would pull them up into a standing position and give them oxygen to breathe. One diver made it as long as one minute and fifteen seconds, but all were out of breath and reported headaches well into the evening. In retrospect, it was a dangerous experiment, but these crude studies were part of what it was like to be an early rebreather user.

It is hard to imagine a diver forgetting to install their scrubber, but numerous incidents have been reported. I've personally known three people who did not install their scrubbers (and obviously did not do a pre-dive check either). All were successfully rescued due to the quick action of diving partners.

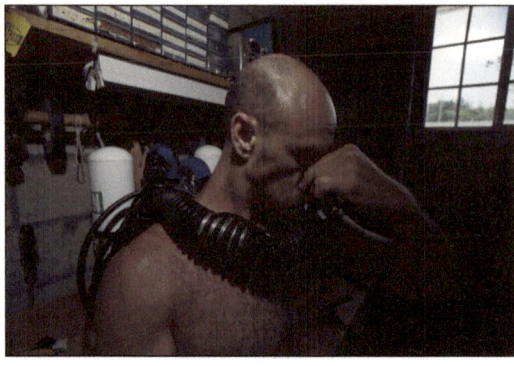

Dr. Kenny Broad completes his pre-breathe with his nose blocked. This prevents the brain from "sneaking" fresh gas in the event of a carbon dioxide issue.

The reaction zone inside a scrubber is unique to each individual design. Gas is routed through the material to create a predictable moving front. Unused sorb is cooler and expended sorb also cools off. Some rebreathers employ an array of thermistors inside the canister to measure temperature through the absorbent and thus predict which sorb is used, active and remaining ready for use.

SAFETY CHECK

Pushing a scrubber with a high workload near the end of the predicted useful range has caused death.

One of the best learning resources regarding carbon dioxide poisoning has been made available by the Health and Safety Executive (HSE) in England. A video showing real footage of a Sky News diver illustrates an incident from a successful rescue that could have easily ended in death. This footage has been made available without copyright in the hopes of preventing similar issues.

COOL

WARM

WARMER

HOT

COOL

CHAPTER 6 - PHYSIOLOGY

There are numerous lessons that can be gained from viewing this short film, not the least of which are team issues and the startling amount of bailout gas needed by a diver with hypercapnia.

Watch: *http://ww.hse.gov.uk/diving/video/co2video.htm*

Several links also exist on the Internet showing helmet camera footage from Dave Shaw's terminal dive in Bushmansgat Cave in South Africa. As grim as it is to watch a diver's dying breaths, one can note his respiratory rate on this high exertion deep dive that likely concluded with with a hypercapnia blackout and drowning.

Carbon Monoxide

Carbon monoxide poisoning is a recognized hazard of diving which results from contaminated air in a SCUBA tank. This contamination may be caused by poor compressor maintenance or poorly placed compressor intakes that are inadvertently contaminated by engine exhaust. Although it is considered rare, I've personally known four divers who died from carbon monoxide poisoning. Though none were on rebreathers, all divers are at risk.

The CO-Pro Personal Carbon Monoxide Protection device is a balloon equipped with a CO detector. It can be used on several tanks before discarding.

In many parts of the world, frequent air testing is commonplace. When you travel, the dive shop may not be as reputable. However, even U.S. fill stations have been responsible for filling tanks with lethal doses of carbon monoxide. Carbon monoxide analyzers are now readily available to divers. Operating similarly to an oxygen analyzer, these devices are portable and economical. An even cheaper and lighter method of analysis has been developed by Lawrence Factor. Their product, known as a CO-Pro, is a small condom-like balloon with sensor material packed inside. When traveling, several divers can use this disposable and inexpensive device to check if their gas is within safe tolerances of carbon monoxide. The condom is slipped over the tank valve and filled from the tank. Within a few minutes, you can observe whether a color change has occurred on the sensing material, indicating a problem with the gas in the tank.

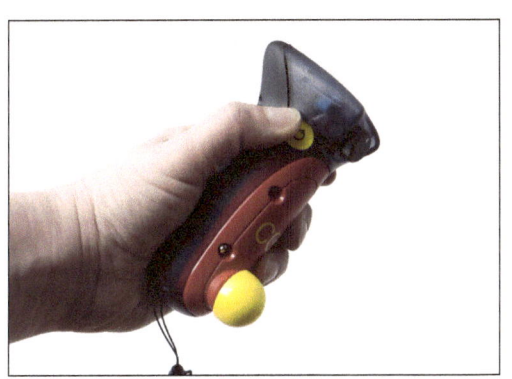

The Analox Carbon Monoxide analyzer is portable and easy to calibrate with your own breath or calibrated bump gas.

107

CHAPTER 6 - PHYSIOLOGY

Another source of carbon monoxide, is from the diver himself. Smokers retain high partial pressures of carbon monoxide at the cellular level after smoking. Carbon monoxide levels of 10 parts per million (ppm) or higher are considered hazardous to the human body. Yet, a study published by W.R. Grace & Company in 1993,[6] revealed that levels as high as 810 ppm have been measured in some anesthesia breathing loops. In the study, they theorize that the carbon monoxide was released from the hemoglobin molecules well after smoking. In a dive profile, those molecules will continue to circulate through the breathing loop until the diver ascends. The bottom line is that smoking and rebreathers simply do not mix. If you smoke, it is definitely time to take action to quit.

Chemical Injury

If you flood your breathing loop, excess water may enter the scrubber canister and react with the absorbent material. Luckily, these "caustic cocktails" are rare, but understanding the basic chemistry will help you respond correctly. If you have ever wet your fingers and handled rebreather absorbent, you may have felt a soapy feeling on your fingertips. This reaction is actually the highly alkaline byproduct of mixing sorb with water removing the top layer of your skin. Because caustic cocktails are alkaline, not acidic, one of the best ways to reverse the alkalinity is to neutralize with a mild acid. Coca-Cola or an acidic fruit juice may help stop the "burn." However, caustic cocktails may also cause uncontrollable coughing. You must not provide liquids to someone suffering from laryngiospasms, breathing difficulty or vomiting.

John Dalla-Zuanna carefully building up the loop of his sidemount rebreather. Attentive preparation combined with negative and positive leak tests lessen the likelihood that water will find its way into the absorbent material and create a caustic cocktail.

Your rebreather loop can become flooded for one of many reasons:

* Improper assembly
* Loss of mouthpiece
* Leaking mouthpiece
* Poorly fitted mouthpiece
* Dropping the loop from your mouth while diving

CHAPTER 6 - PHYSIOLOGY

* Torn loop hose
* Damaged counterlung
* O-ring failure in the DSV, CCR body or hose connections
* Failure to completely move loop lever into closed-loop position
* Failure to close loop at surface
* Failure to install scrubber drain plug (in some models)
* Failure to close scrubber drain after draining (in some models)

Chemical injury can also happen when scrubber material is not handled properly on the surface. Gloves, mask and eyeglasses will prevent getting active scrubber dust into moist areas of the body such as the eyes and airway.

Decompression Issues

The partial pressure of inert gas determines the decompression obligation for a diver. With open circuit and SCR diving, the maximum PO_2 is reached at the maximum depth of the dive. Hybrid or Intelligent eSCRs utilize a floating setpoint that allows for slightly more decompression efficiency. Closed circuit rebreathers offer the best decompression efficiency because they can be set to a constant PO_2 (setpoint) that minimizes decompression. That doesn't mean that CCRs are safer in terms of decompression. Any device can be dived at its limits, and divers have proven that they are using CCRs for increased range. Safe decompression planning will be covered in the Chapter Seven.

Safety diver Brett Gonzalez (right) drops in on Paul Heinerth (left) after a deep dive off the Bermuda Bank. Gonzalez takes sample bags and additional deep bailout tanks that Heinerth will no longer need to complete his long decompression.

Narcosis

As with open circuit diving, inert gas narcosis can affect a rebreather diver too. Symptoms may include motor difficulties, difficulty with reasoning, inability to handle stress, panic, perceptual narrowing, visual field narrowing, euphoria and anxiety. Divers often associate gas narcosis with diving too deep on air. This is certainly the most common cause, but rebreathers offer a new way to get "narked."

I recall diving with my partner in a deep cave. We both carried appropriate gas for the dive. The rebreathers we were using required that we begin the dive with a 0.4 setpoint. We plummeted to the depths, arriving at a PO_2 of just over 1.3. At

CHAPTER 6 - PHYSIOLOGY

depth we dropped some bailout tanks and switched the setpoint up to 1.2. Well, at least I did. My partner left his setpoint at 0.4 and continued into the cave. When we reached our turn point, he signaled for me to look at his handset which indicated a severe decompression obligation. It was at this moment that I noticed that his setpoint was still set at 0.4. Worse yet, his PO_2 was the same. Throughout the inbound portion of the dive, his actual PO_2 had been eroding until the solenoid saved him and at least maintained a 0.4. At 300 feet this wasn't just a decompression issue. I realized he was absolutely narked! The partial pressure of inert gas in his loop was far higher than normal and the nitrogen in his trimix gave his head quite a wild ride. On his trimix diluent gas of 18/33, he had actually netted an Equivalent Narcotic Depth around 250 feet. The dive ended uneventfully after we switched his setpoint, but I realized the problem could have been far worse. All we had to do was some extra decompression time.

When you are well trained and practiced, operating your rebreather should be easy and relaxed. If it ever feels like diving is a struggle, then step back to a simpler dive until operations become intuitive again. Lynn and Bruce Partridge of Shearwater Research (above) have applied simplicity to the user interface in their Shearwater line of computers.

As previously mentioned, you are carrying a gas mixing station on your back. There are many ways you can screw up the mix that is supposed to be your life support. Your rebreather needs to become a physiological extension of your body. More than any other diving equipment you have owned before, you need to create a harmonious rhythm of monitoring and operations that makes for a smoothly controlled dive.

---

[5] Butler, Frank K.; White, E.; Twa, M. (1999). "Hyperoxic myopia in a closed-circuit mixed-gas SCUBA diver". Undersea and Hyperbaric Medicine 26 (1): 41–5. PMID 10353183. Nichols, C.W.; Lambertsen Christian (1969). "Effects of high oxygen pressures on the eye". New England Journal of Medicine 281 (1): 25–30.
Shykoff, Barbara E. (2005). "Repeated Six-Hour Dives 1.35 ATM Oxygen Partial Pressure". Nedu-Tr-05-20 (Panama City, FL, USA: US Naval Experimental Diving Unit Technical Report).
[6] Sodasorb Manual of CO2 Absorption, W.R. Grace & Company, 1993, section M-18

7

Procedures

In this chapter:

- *Accident Analysis*
- *Dive Logistics*
- *Oxygen Planning*
- *Choosing the Best Mix*
- *Time, Depth and Distance Planning*
- *Decompression Theory and Procedures*
- *Packing the Canister*
- *Pre-dive Checks*
- *In-water Procedures*
- *Mixed Teams*

Accident Analysis

A diver plunges into the ocean wearing his high-dollar technical rebreather, excited to get to the wreck below. There is only one problem: he failed to turn on his gas supply. Meanwhile, across the globe another diver has spent her hard-earned money to get out to a rarely accessed site. She readies her rebreather and finds that one of her oxygen sensors is acting strange. She decides to dive with her rebreather anyway. A third diver, worlds apart, prepares for a deep decompression dive. He has wisely chosen to purchase a rebreather to conduct this difficult dive. He has foolishly decided to forego carrying or staging adequate backup gas supplies, reasoning that a "lean and mean" assault on the site is a better idea. An abort, a rescue and a fatality occurred as a result of the above actions by divers. It doesn't matter who was involved or which was the worst error. Any one incident could have resulted in death.

These are real scenarios that have occurred in our community over the past several years. As the popularity of rebreathers increases, the number of incidents and accidents are also on the rise. But as demonstrated through the events above, a culture of safety has not yet evolved. These days, divers can walk into a dive shop with a credit card and walk out with enough gear to get them to the end of almost any cave line on the planet. They can reach new ocean depths and stay for longer than we could have imagined before. In some ways, this re-

CHAPTER 7 - PROCEDURES

minds us of the genesis of cave diving in Florida. It was a frontier filled with opportunity and very few rules of engagement.

In the 1970s, cave diving pioneer Sheck Exley began to recognize patterns in the many fatalities that were occurring in caves. As a result, he authored the Rules of Accident Analysis that have helped to educate divers since that time. In his Blueprint for Survival[7], he carefully analyzed cave diving incidents and fatalities and categorized errors into a list of Golden Rules for diving in the overhead environment.

The technical diving community lacks a similar code of conduct for rebreather diving. With rebreather companies and models changing about as quickly as Microsoft Windows upgrades, it is difficult to address. However, in view of increased rebreather diving activity, it is a timely and important discussion. Even considering the operational differences of various models, a basic list can be constructed.

Training

Several fatalities on rebreathers have been attributed to lack of training and proficiency. You need ample training on your unit and well-practiced situational emergency skills. Considerable water time is needed to hone the intuitive dexterity that is critical for rebreather divers. The equipment is a physiological extension of your body. You are truly integrated into the breathing loop and need time and experience to develop an intuitive link with your gear. If you are struggling with stress from taskloading, you might disregard a change in the pace of your solenoid valve or fail to question "that odd feeling" you are experiencing as a result of high carbon dioxide or low oxygen levels in your loop.

Rebreather training is unit specific. Each time you start diving a new rebreather, you will need training in unit specific skills.

Pre-Dive Checks

If you are not meticulous and diligent with pre-dive checks then you shouldn't consider owning a rebreather. Most manufacturers have developed stringent guidelines for pre-dive preparation of their rebreather. Generate a log page with a checklist that you can use every time you prepare your rebreather for a dive. A written or digital checklist ensures that critical safety issues are never overlooked because of a momentary distraction. It also serves as a record of consumables and a diagnostic log that may aid in identifying developing problems with your gear.

CHAPTER 7 - PROCEDURES

A proper pre-dive check must include a pre-breathe sequence of five minutes, with the nose blocked. This should be done seated, out of the water, while watching displays. A pre-breathe of this duration will catch most errors such as carbon dioxide breakthrough, inoperative solenoid valves or failing to turn on your gas supply.

Decision to Dive

Pre-dive checks need to be conducted in an unhurried manner and without the pressure of rushing to jump in the water. Rebreathers require more care and maintenance than open circuit gear. It is critical that you heed the warnings of a failed safety check. I have witnessed numerous individuals who have elected to dive with an errant sensor or a "small" leak. If you decide to jump into the water with a system error, you might as well have one foot in the grave. There is no dive that is worth the risk of starting with half of your options already used up. As mentioned earlier, complacency tends to creep into the routine of experienced divers who have logged repeated, incident-free dives. That same complacency has been a contributing factor in numerous fatalities to date.

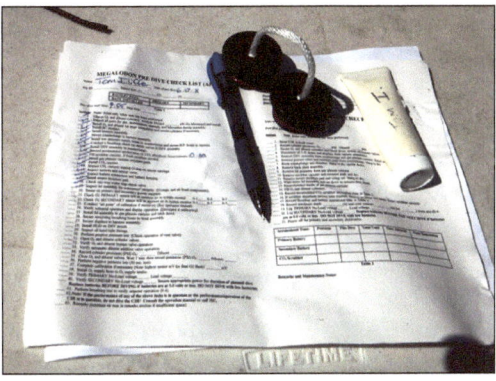

A pre-dive checklist should be used diligently every time you dive. If all systems do not pass the checklist, then you must abort your dive plan completely.

Fitness

Since a rebreather is integrally linked with your physiology, it is crucial that you are in optimum physical and mental fitness for diving. Mental acuity may help you recognize symptoms of a developing problem. If you feel anything unusual, you must be patterned to switch to open circuit or flush with appropriate fresh diluent to determine whether you have a problem. If the problem does not have an immediate solution, then you should stay on open

CHAPTER 7 - PROCEDURES

circuit and get yourself to the surface safely. You should only return to the loop if you are using a technical rebreather and have diagnosed and solved the problem fully. Recreational divers must *always* abort to open circuit.

SAFETY CHECK
If in doubt, bailout!

Contingency

Backup planning is the lengthiest part of organizing a safe rebreather dive. It can be broken down into the following steps that will be covered in further detail in this chapter.

Gas Planning: Plan your onboard supply to accommodate all the depth changes you may experience during your dive. Frequent ups and downs will require more diluent than you might normally anticipate.

Some rebreathers built today have significant onboard open circuit capacity, and a built-in second-stage suitable for open water diving. However, divers should not consider diving below 20 meters/60 feet or entering an overhead environment without adequate bailout gas to reach safety.

Plan your open circuit bailout gas supply to accommodate the worst-case scenario of a catastrophic loop failure requiring an open circuit bailout all the way to the surface. This gas supply can be cached as stage bottles or carried. The "Alpine" approach to rebreather diving that fails to plan for catastrophic failures is foolhardy at best. A rebreather is not an excuse to cut corners in safety. If you cannot independently handle the gas needed for open circuit bailout and decompression, then you should incorporate support divers that can aid in staging an emergency gas supply. The only alternative to this approach is using a fully redundant dual rebreather with redundant gas supply and loop.

When you have been away from diving your rebreather, you should get back into diving slowly. Practicing in a swimming pool or confined water environment will improve your motor skills, reaction time and confidence.

Decompression Planning: If your rebreather or decompression planning tool is electronic, then you must be prepared with alternate plans for decompression. You should carry a redundant backup computer or tables for open circuit bailout and constant PO_2 tables.

Rehearsal: You should practice critical skills for dealing with all possible failures. If you have an electronic rebreather, you should be comfortable with manual control in the event of a full

CHAPTER 7 - PROCEDURES

loss of power. If you have a rebreather with a solenoid valve, you need to get in tune with the sound and pace of that valve and be ready to deal with all the possible failures associated with it. If there is any chance that a loop flood could cause a caustic cocktail, you need to consider how to deal with open circuit bailout, choking and alkaline burns all at once.

Relaxed team meetings are a key component to good dive planning.

Team Planning: Dive buddies and support divers need to be educated about how to recognize and handle emergencies that may be specific to your rebreather. Open circuit divers need to be instructed about how they can access your gas in an out-of-air emergency. They need to be shown what your alarms indicate and how to switch to open circuit bailout. They need to know how to close your loop in case of unconsciousness and need for rescue. They need to plan their contingencies with consideration of the amount and type of open circuit gas that you are carrying.

If team members intend to share the responsibility of providing adequate open circuit bailout gas - a technique I do not personally endorse except on extreme expeditionary profiles - then consideration must be given to different buoyancy and clip systems on each other's cylinders. A rebreather diver that has to accept a tank that is out of trim may be jeopardized from a quick abort.

When the chips are down, the features of your rebreather are only as good as your ability to solve problems creatively and survive with or without those features. Make a set of laminated cue cards with all possible failure modes that your rebreather can present. On an open water dive, use the cue cards with your dive buddy to rehearse each emergency as a team. If you dive with an open circuit buddy, ensure that they know how to assist with any of those eventualities.

Rebreathers give us a gift of time to solve most problems. But it is unwise to push the learning curve too quickly and task load to the point of serious stress.

Dive Logistics

Gas Planning and Fills

One of the first things to figure out when you are planning a dive is how much gas you consume on an ideal dive. We looked at onboard tanks in the physics section of the book, but what about bailout gas? How much is enough? As an open circuit diver, you may be familiar with the concept of Surface Air Consumption (SAC) Rate. You can't simply take your SAC Rate from open circuit diving,

CHAPTER 7 - PROCEDURES

since it may be different if your rebreather is bulkier and less streamlined. Technical divers calculate a SAC Rate to help them compare air consumption with other team members and determine the range for their dive. Rebreather divers need to know how much gas is necessary to get them to the surface with a safe margin of error in the worst case scenario. To calculate this figure, you can swim for at least 10 minutes at a given, stable depth in a full set of rebreather equipment. Walk through the exercise below to help translate your air consumption to an equivalent surface figure in the following way:

Calculating your SAC Rate

IMPERIAL:

$$SAC\ Rate\ (cft/m) = \left[\frac{[psi\ used \div working\ pressure\ (psi)] \times cylinder\ capacity}{(Depth\ (ft.)+33) \div 33}\right] \div minutes$$

METRIC:

$$SAC\ Rate\ (lpm) = \left[\frac{bar\ used \times total\ cylinder\ capacity\ (ltrs)}{(Depth\ (m) + 10) \div 10}\right] \div minutes$$

Calculated in this manner, the SAC rate is an ideal figure that must be padded for less than ideal scenarios. Some divers calculate their SAC rate under high workload so they have a range of numbers that can be used in dive planning.

Tank Baseline

You should consider how much bailout gas your buddy is carrying. If it looks lean, then he will be looking to you for more if he needs it. If he is carrying a different size tank filled to a different pressure, then there is a little math required to figure out how much gas he is carrying.

When tanks are filled, they rarely get filled precisely to their working pressure. If using the Imperial measurement system, you should compare gas volumes between different members of the dive team. After the tank has been filled, you can use a number called a "tank baseline" to compare volumes between different fills and different sizes of tanks. The baseline describes how many cubic feet are represented by one psi on the diver's gauge.

Tank Baseline = Cubic Foot Capacity / Tank Working Pressure

If Diver A uses an LP 104, he would calculate:

Baseline = 104 / 2640 = .039 cu. ft. per psi

If Diver B used double aluminum 80s, she would calculate:

80 / 3000 = .027 cu. ft. per psi

CHAPTER 7 - PROCEDURES

To calculate the total volume in a diver's tank, simply multiply the tank baseline by the actual fill pressure.

For example:

If Diver A has a 104 cu. ft. steel tank with 2500 psi, he would calculate:

.039 x 2500 = 97.5 cu. ft. of gas

If Diver B has an 80 cft. aluminum tank with 2500 psi in the tanks:

.027 x 2500 = 67.5 cu. ft. of gas

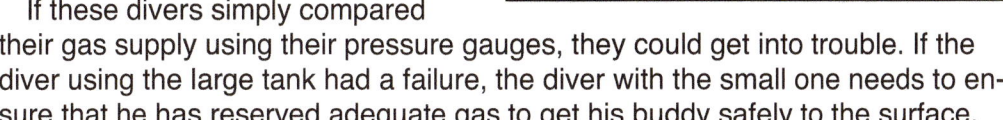

If these divers simply compared their gas supply using their pressure gauges, they could get into trouble. If the diver using the large tank had a failure, the diver with the small one needs to ensure that he has reserved adequate gas to get his buddy safely to the surface.

SAC Rate is used to determine how much gas volume you need to get home, but you should also think about what particular issues might cause you to abort to open circuit. In recreational rebreather diving, the only option in an emergency is to abort to open circuit and ascend directly to the surface, with a safety stop if possible. In technical diving, direct ascent may not be possible, so you need to think about every step of an abort process which can include a long exit from a wreck or cave or a lengthy decompression (which acts as an artificial overhead environment). Fortunately, few rebreather issues will result in the necessity for open circuit bailout from maximum penetration or dive time. These issues are referred to as Catastrophic Loop Failures. In other words, you cannot use the rebreather at all. Some catastrophic loop failures include:

* Hypercapnia (failed scrubber)
* Flooded loop
* Lost loop integrity (ripped or torn loop hoses, broken canister, etc.)

SAFETY CHECK

Diving with an open circuit partner means you need to carry adequate bailout to get them safely home in an emergency situation.

As long as your loop is intact and you still have diluent gas, then a technical diver has options. A rebreather is simply a big bag of gas. If your diluent gas is safe to breathe at max depth, then you can always flood the loop with diluent to create a safe, predictable gas mix as long as you know your depth. If you flush your loop with diluent, then you have the luxury of time to figure out what to do next. During a technical rebreather class, you will learn how to operate a CCR in SCR mode by exhaling every "nth" breath and re-flushing with diluent gas to replace the volume. In this mode, the diver can extend their gas supply substantially, even if they do not have working electronics or sensors. If you still have

CHAPTER 7 - PROCEDURES

oxygen left, you have even further options. At a stable depth, you will learn to operate the rebreather with a minimal loop volume, injecting oxygen only when the loop volume drops. If you do not lose volume in other ways by venting or bubbling through a mask, then you can maintain a constant PO_2 very effectively, even without electronics.

However, if you are forced off the loop from something such as hypercapnia, then you will need a tremendous amount of gas. A diver with a high CO_2 in their bloodstream will be breathing at a substantially increased respiratory rate. I once rescued a diver with hypercapnia from a cave and was astounded by the quantity of gas needed to get him to the surface but also by his inability to slow his breathing rate. It does not matter how experienced you are, if you are having a carbon dioxide emergency you may not be able to slow your breathing rate.

As a result of these observations, I choose to carry a healthy amount of bailout gas and split it between two separate tanks. This balances my trim and offers increased redundancy. If my buddy ever needs open circuit gas from me, I can pass an entire tank without unhooking my dry suit inflator or being left without gas for myself.

Of course, I would be lying if I told you that I would never consider sharing bailout responsibilities with a team. There are some extreme expeditions where I have considered staged gas as community gas. I use my cutoff point as swimming with two to four full-sized cylinders, depending on how much swimming or scootering will be required. Beyond that, I will share the rest of the supply. Those two to four tanks should be able to get me a considerable way towards a safe exit. In

Preferring to carry weight as gas rather than as lead, the author's standard bailout configuration for simple dives is to use 2 x 45 cft. steel tanks with her rebreather. Photos: Mark Long.

CHAPTER 7 - PROCEDURES

If you are adequately covered for contingencies, then your rebreather kit will not be any lighter than a deep open circuit setup. The deep rebreather kit offers you far more options in the event of emergency as long as you have planned ahead with adequate bailout gas.

regular sport dives- average to verging on extreme- I see no reason to share bailout gas.

Some training agencies require that a team of three divers working together carries enough gas to get 1.5 divers home safely. In other words, take the highest SAC Rate of the team, multiply that by 1.5 and then divide that gas amongst the team of three. In my personal assessment, this is too risky. Not one of the divers on that team will have enough gas to get home if left alone. If the team stays together, then it will be necessary to complete successful tank swaps during abort. In my opinion, that is a lot to ask of an already stressed group. Tank swaps are conducted when a tank is half empty so that nobody is left without an open circuit gas supply. That means several tank swaps may be needed if this emergency happens at the worst possible time. I know the likelihood is low, but why add to your risk profile when you can easily mitigate the risk? It is better to carry gas you don't need than to need gas you don't have.

Oxygen Planning

Choosing the Best Diluent and Bailout Gas

It is important to plan for oxygen exposures when selecting an appropriate gas and setpoint for your dive. In the past, divers commonly used a partial pressure of 1.4 for the maximum operating depth of a given gas and 1.6 as the maximum PO_2 for decompression gas. Most technical divers, and especially cave divers, have chosen to reduce their oxygen exposure for both safety and extending the available oxygen clock. Many technical training agencies now suggest 1.3 as the maximum PO_2 exposure for the open circuit bottom time and 1.4 to 1.6 for decompression, when the diver is at rest. CCR divers commonly use 1.2 or lower for their setpoint. As mentioned in the chapter addressing physiology, oxygen planning needs to account for central nervous system (CNS) Toxicity and Pulmonary or whole body toxicity.

CHAPTER 7 - PROCEDURES

The following oxygen exposure limits have been determined by NOAA for tracking CNS oxygen toxicity:

NOAA OXYGEN EXPOSURE LIMITS

Oxygen Partial Pressure	Max. Exposure for Single Dive	Max. Exposure for 24 Hours
0.6	720 minutes	720 minutes
0.7	570 minutes	570 minutes
0.8	450 minutes	450 minutes
0.9	360 minutes	360 minutes
1.0	300 minutes	300 minutes
1.1	240 minutes	270 minutes
1.2	210 minutes	240 minutes
1.3	180 minutes	210 minutes
1.4	150 minutes	180 minutes
1.5	120 minutes	180 minutes
1.6	45 minutes	150 minutes

Calculate the percentage of oxygen toxicity (also known as the oxygen clock) by looking up the partial pressure at maximum depth, or rebreather setpoint, in the table above. Then divide 100 by the number in the second column to find the percentage per minute. Finally, multiply that number by the total dive time. This will give you a percentage of the oxygen clock.

CNS % PER MINUTE CHART

PO_2	CNS % per Minute	PO_2	CNS % per Minute
0.7	0.18	1.2	0.48
0.8	0.22	1.3	0.56
0.9	0.28	1.4	0.67
1.0	0.33	1.5	0.83
1.1	0.42	1.6	2.22

CHAPTER 7 - PROCEDURES

You may prefer to use the chart on the lower part of the previous page, that gives you a *p*er minute exposure based on various partial pressures. Use the partial pressure at maximum depth or rebreather setpoint and multiply by the right column to come up with the percentage of the oxygen clock.

Residual Oxygen Toxicity

After a long surface interval, you can calculate the residual percentage of CNS oxygen toxicity. After a 90 minute half-time (surface interval), your oxygen toxicity percentage will be halved. The chart below makes it simple to look up.

CNS TOXICITY HALF-TIME CHART

Surface Interval Time

CNS %	90 Min	120 Min	150 Min	180 Min	210 Min	240 Min	300 Min	360 Min
100	50	38	30	25	22	19	15	13
95	48	36	29	24	21	18	15	12
90	45	34	27	23	20	17	14	12
85	43	32	26	22	19	16	13	11
80	40	30	24	20	18	15	12	10
75	38	29	23	19	17	15	12	10
70	35	27	21	18	15	14	11	9
65	33	25	20	17	14	13	10	9
60	30	23	18	15	13	12	9	8
55	28	21	17	14	12	11	9	7
50	25	19	15	13	11	10	8	7
45	23	17	14	12	10	9	7	6
40	20	15	12	10	9	8	6	5
35	18	14	11	9	8	7	6	5
30	15	12	9	8	7	6	5	4
25	13	10	8	7	6	5	4	4
20	10	8	6	5	5	4	3	2
15	8	6	5	4	4	3	3	2
10	5	4	3	3	3	2	2	2
5	3	2	2	2	2	1	1	1

New CNS Percentage

CHAPTER 7 - PROCEDURES

NOAA has made further recommendations for the minimum surface interval between dives. Details on their recommendations for diving are provided free online using the search term "NOAA Diving Manual."

Calculating Pulmonary Oxygen Toxicity Exposure

In addition to CNS toxicity, rebreather divers need to monitor their exposure to pulmonary/whole body oxygen toxicity. In order to track pulmonary oxygen exposure, Dr. Bill Hamilton developed a method called the "Repex Method," which uses a measurement called an Oxygen Toxicity Unit (OTU). These are calculated on single dives and accumulate after repetitive days of diving. Theoretically, staying within these limits should safeguard the diver from experiencing symptoms of pulmonary toxicity and still allow room for recompression therapy if it is needed.

CALCULATING OTUs USING DR. BILL HAMILTON'S REPEX METHOD

Pulmonary Dose

Partial Pressure	OTU per Minute
0.6	0.27
0.7	0.47
0.8	0.65
0.9	0.83
1.0	1.00
1.1	1.16
1.2	1.32
1.3	1.48
1.4	1.63
1.5	1.78
1.6	1.92
1.7	2.07

Daily Allowances

Days	Daily Dose	Total Dose
1	850	850
2	700	1400
3	620	1860
4	525	2100
5	460	2300
6	420	2520
7	380	2660
8	350	2800
9	330	2970
10	310	3100
11	300	3300
12	300	3600

CHAPTER 7 - PROCEDURES

Choosing the Best Mix

The mix in your diluent bottle should be safe to breathe at maximum depth. It should prevent narcosis and be capable of flushing the breathing loop both up or down from dangerous PO_2. Some instructors will recommend that only air is put in a diluent bottle, but you may have a reason to use a different gas. When working with inexperienced Hollywood actors on a shallow underwater set, I chose to turn the production rebreathers into oxygen rebreathers by filling both tanks with pure oxygen. The gas was safe to breathe in the ten foot deep pool and there was no way anyone could experience hypoxia or hyperoxia. It made sense. Similarly, there is no reason why a diver could not put a moderate nitrox mix in the diluent tank of a shallow dive as long as it was breathable at maximum planned depth.

Now that mixed gas and appropriate training for deep diving are readily available, most divers shy away from deep air diving and the narcosis that accompanies it. Deep diving is an area of rebreather diving where you will see a significant cost savings over open circuit deep diving. In deep applications, most rebreather divers choose to operate with a maximum Equivalent Narcotic Depth of between 100 - 130 feet, with ever more divers choosing the shallow end of the spectrum. This facilitates clear thinking and good problem management capability in the most difficult scenarios. Even though a diver may believe that they are not impaired, they will experience greater challenges in task loaded scenarios when diving high partial pressures of nitrogen.

The EAD (Equivalent Air Depth) formula gives divers the ability to compare a given mix of gas to a theoretical depth equivalent for diving air. This figure may be used with air diving tables. In a rebreather loop, the proportion of oxygen to nitrogen is generally great than in air. Therefore your EAD is less than your actual depth. The EAD fluctuates through the dive as the depth or setpoint changes. (See the Physics Chapter for this formula.)

Calculating the Best Mix (based on your choice of PO_2 at maximum operating depth)

ATA = pressure at maximum operating depth

Chosen PO_2 = this is the desired PO_2 of the diluent tank, which in an ideal world, would be within .1 or .2 of your planned setpoint. If you are planning a setpoint of 1.2, then a diluent bottle filled to 1.0 at MOD would be ideal.

$$FO2 = \frac{Chosen\ PO2}{ATA}$$

You may still choose to fill you diluent tank with air on most dives, but think about the following situations:

CHAPTER 7 - PROCEDURES

✓ If I breathe from the BOV and have a large tank attached to the BOV, will the diluent PO_2 be suitable for bailout and getting to the surface reasonably quickly?

✓ If I have a hypoxic loop and flush with diluent, will that diluent gas bring the loop PO_2 up to a life sustaining level?

✓ If I have a hyperoxic loop from descending too quickly (or another reason), and I flush with diluent gas, will that gas bring the loop PO_2 down in a satisfactory manner?

✓ If I am uncertain of the accuracy of my oxygen sensors at maximum depth and I flush with diluent gas, will I be able to recognize a change in PO_2 and decide which sensor(s) are operating properly?

✓ What PO_2 should I see on my display if I flush with diluent at maximum depth?

✓ Will my best mix for deep diving be hypoxic in shallow water through the BOV?

Time, Depth and Distance Planning

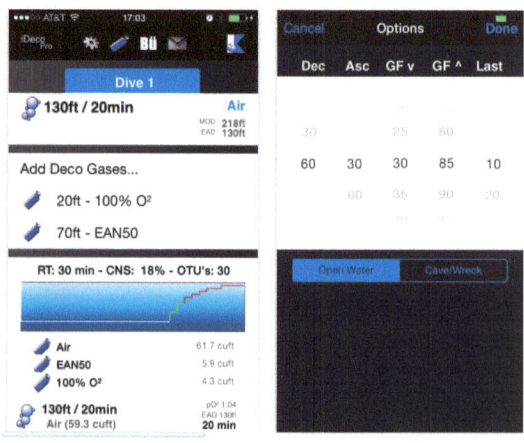

There are several factors that come in to play when choosing your dive depth, duration and possibly penetration of an overhead environment. It all comes down to your consumables and desired decompression exposure. Using your SAC rate, you can determine a projected abort range in the event you have to complete a dive off the loop. By taking your SAC rate and multiplying it by the depth in ATA/bar, you can project potential gas usage at depth. The range of a particular cylinder can be calculated by taking the volume of the fill and dividing it by the SAC rate at depth which is sometimes referred to as the respiratory minute volume (RMV). This calculation only highlights the potential range of a tank and does not add any conservatism or account for differing volume in a partner's bailout tank. These calculations can be made for every stage of an ascent or can be conservatively generalized by using maximum depth.

Decompression planning software such as V-Planner or iDecoPro will do these calculations for you when you input your SAC rate and available gases. Use of software is highly recommended to avoid calculation errors. The software is used for dive planning, but backup decompression computers are most often used for dive execution. Some divers use tables as a backup, but they are becoming more rare as computers become less expensive and more features get built into onboard secondary systems. The most advanced rebreathers are equipped with simulators and decompression diving software packages for dive planning.

CHAPTER 7 - PROCEDURES

There are other factors which may require you to plan a shorter dive or abort a planned dive. They include:

* low onboard oxygen supply
* low onboard diluent supply
* low remaining scrubber time
* low remaining battery time
* oxygen exposure limits
* thermal considerations
* desired maximum decompression time
* environmental factors such as tides
* system failures such as a bad sensor
* other emergencies

Thermal considerations may outweigh all other factors when planning your turn time for a dive.

A dive is turned at a time that allows you to reach the surface with a solid margin for safety based on any single consumable. Some divers use the "rule of thirds" for calculating this range. They use one third of any given consumable for the beginning portion of the dive, one third for the return and leave one third in reserve for contingency needs. Some recreational rebreathers constantly monitor all consumables and readjust the floating setpoint to take all parameters into account, eventually warning you when supplies reach a turning point. Other rebreathers require you to monitor each consumable independently and turn the dive at an appropriate time.

Decompression Theory and Procedures

If you train on a technical rebreather, you will need to learn some essentials about decompression. Additional in-depth training in decompression procedures is recommended as well as rescue diving, oxygen administration and diving first aid. A complete education in decompression theory is beyond the scope of this text, but some general concepts will aid divers in their understanding of various decompression models that are used today.

When a diver is under pressure, his tissues absorb inert gas. Different tissue types absorb gas at various rates. Relative solubility of gas varies between bone,

CHAPTER 7 - PROCEDURES

fat, brain and other tissues. As such, the body is divided into theoretical tissue groups, known as compartments. The amount of gas dissolved within certain tissues is what controls a decompression model. When a tissue compartment becomes saturated, it reaches a level called the "M-Value." The term M-Value was first described by Robert Workman of the U.S. Navy Experimental Diving Unit (NEDU)[8]. The letter "M" was intended to describe the "maximum" value that a tissue compartment could tolerate, without exhibiting signs of over-pressurization or supersaturation. Understanding the rate at which tissues reach supersaturation, helps mathematicians develop models to predict safe ascent protocols.

Historically, divers relied on tables developed by the military. The U.S. Navy Decompression Dive Tables offered some of the earliest resources to divers. They were designed for single exposures, by young, fit males using air for bottom time and decompression on open circuit dives. This type of diving bears little resemblance to technical rebreather diving today, therefore more suitable models have been developed. New tables and computer software help a diver carrying a variety of gases, to plan profiles that better prevent DCI incidents.

In the earliest models, divers were encouraged to rapidly ascend to a series of shallow decompression stops where they could sufficiently off-gas nitrogen or other inert gases. Research has indicated that a slower ascent strategy with preventive deeper stops tends to help the diver off-gas more efficiently and prevent

the growth of gas phase (an index of bubble radius relative to the initial size of a gas nuclei), rather than treating micro-bubbles in tissues at their point of supersaturation. This strategy of starting decompression deeper may shorten shallow stops, but excessive deep stops can contribute to more inert gas loading.

Models such as the Reduced Gradient Bubble Model (RGBM), Variable Permeability Model (VPM), Variable Bubble

CHAPTER 7 - PROCEDURES

Model (VBM) and Variable Gradient Model (VGM) are popular decompression algorithms for technical divers. These models modify their ascent to a gradual curve with stops beginning deeper. Bühlmann tables have been modified using a concept called Gradient Factors which permit a diver to choose the degree of conservatism in their profile. By determining theoretical M-Values, the model predicts when supersaturation will occur in different tissue compartments. The diver chooses how close they want to get to supersaturation at their first (deepest) stop and how conservative they want the overall profile to be upon exiting the water. Erik Baker, an electrical engineer and computer modeling specialist, offers up many resources on the Internet that fully describe gradient factors in plain, easy to understand language (Search online: "Gradient Factors for Dummies").

Some computers permit the diver to select a personal gradient factor. Several computer manufacturers, such as Shearwater, set a default gradient factor of 30/70. What this means is that the deepest stop will be generated when the controlling tissue compartment reaches an M-Value of 30% of the quota between ambient pressure and acceptable supersaturation. The rest of the stops slowly ease the diver up to 70% of the theoretical supersaturation level, in essence setting an exit conservatism of 30%, but with gradual off-gassing along the way.

In the Winter 2010 issue of *Alert Diver,* several noted experts weighed in on their opinions of deep stops. Generally, all agreed that computers and table models that already incorporate deep stops may be safely used by divers. They argue, however, that any arbitrary adjustment of any diving algorithm may be very risky. Christian Gutvik, noted Norwegian researcher, reminds us that, "our current theoretical models and experimental results indicate that deep stops are beneficial only on longer dives."

Many CCRs and some SCRs are equipped with onboard decompression computers that calculate decompression on the fly based on the actual PO_2 in the loop. Depending on the algorithm in your computer, you may be able to set a conservatism factor or change the gradient factors to suit your desired risk profile or match another rebreather diver or backup computer. Some computers will allow you to access several different decompression models, through downloads from a web site. A few rebreathers allow the diver to completely alter the deep, mid, and shallow stop conservatism or aggression. Great care must be taken in this instance, since you will be using an untested algorithm mix.

CHAPTER 7 - PROCEDURES

Specific training in decompression procedures is available from most agencies. This training will help you refine skills and give you background knowledge for handling oxygen-rich mixes and calculating decompression plans. If you are taking a technical rebreather program, this will be handled in your class. If you are taking a recreational program, you may receive training in handling high FO_2 mixes, but not necessarily any specific instruction in decompression procedures and models.

As a technical rebreather diver, you will always have oxygen available on site inside your rebreather, but whenever decompression dives are planned, consider having additional oxygen available for emergencies. After any decompression dive, it is wise to spend relaxed time on the surface if thermal conditions permit. If this is not possible, then it is prudent to spend some low exertion time relaxing after your dive. This is called "surface decompression." At the end of your dive you are often breathing a PO_2 between 1.2 and 1.6 at 20 feet / 6 meters. Off-gasing is very effective with this elevated PO_2. Immediately upon surfacing, your PO_2 drops to .21 and off-gasing slows. You should avoid strenuous exits if possible. Lugging a heavy rebreather immediately up a set of stairs, or carrying bailout bottles out of the water may require heavy exertion. This extra workload could mean the difference between remaining asymptomatic and getting the bends. Surface decompression gives the body a chance to continue effective off-gasing prior to undertaking any heavy workload. In general, surface decompression time should be at least half as long as the decompression time required at the last stop.

You may be fortunate enough to have an ascent line for decompression, but rebreathers divers will often shoot a lift bag (DSMB) to notify the vessel of their location and status.

CHAPTER 7 - PROCEDURES

A rebreather uses no more gas on decompression than it does at depth, as long as the operator is diving it effectively. Many eCCR divers find that the most effective way to minimize gas usage is to operate their unit manually during ascent and decompression. This aids in gas conservation by allowing you to prepare ahead for your next stop. As you leave a stop, exhale volume from the breathing loop in anticipation of gas expansion caused by ascent. Leaving a reasonably high setpoint (1.0) on the controller ensures a safety net, but practice getting a little ahead of the solenoid by keeping the PO_2 slightly higher than setpoint. If you don't operate in this manner you may find that you are venting a lot more gas than necessary as the solenoid overdrives the volume in the loop and must be needlessly dumped before it gets fully homogenized. Ascending through decompression stops effectively is a real dance, where timing is critical every step of the way. You'll practice this in your class and your instructor will give you specific tips for the brand of rebreather that you have purchased.

If you are completely new to decompression diving during your technical rebreather class, ensure that your instructor knows this. It is important that you are also well versed in open circuit decompression and gas switch techniques. If your buddy makes an open circuit gas switch, you should watch them and carefully trace the open circuit regulator back to the tank to confirm the appropriate contents marking and depth at which the switch is made. If a safety tank is left on a drop line or in a cave/wreck entrance, the tank should be clipped to the line and pressurized, but turned off until it is needed.

Your rebreather may be equipped with a HUD that notifies you about your decompression status and when it is time to move to the next stop. If your brand has this capability, you will learn to conduct HUD-only decompression in your rebreather class.

Stops should be made as close to the required depth as possible, but bear in mind that your partner's profile can vary from yours. Your actual PO_2 may have been different than your buddy's during the dive, netting a slightly different decompression. Rebreathers give you a luxury of time that should permit you to stay together through the most conservative profile.

CHAPTER 7 - PROCEDURES

In some cases you will be able to descend and ascend on an anchor or mooring buoy. In open water, DSMBs are often used for decompression and some no-decompression ascents. The marker buoy marks your position and let's the boat crew know you have reached a pre-determined stop depth. In some countries, different colored markers are used for "okay" or "need help." In some cases, such as some drift diving applications, everyone will shoot their own lift bag, but in other situations it may only be necessary to shoot one for a team of divers. You will always carry your own lift bag and spool/reel for deployment in case you need it. You'll learn how to do this in closed circuit and open circuit mode during your class.

When your team is discussing their decompression strategy, you should organize who will send up a lift bag, how you will stay together and how you will determine when it is time to call the dive. If one diver generates a deep stop, will the entire team stay at the stop or just within visual distance? If one diver clears decompression obligations first, will they stay until the entire team has cleared? Although I never begrudge someone getting out of the water early because they are cold, I would never leave a team member alone on decompression.

Comparison of Various Decompression Algorithms

Manufacturers use various decompression algorithms in their rebreathers and computers. Your backup computer may contain a different algorithm from your rebreather. The following is a chart of numerous current rebreathers and the algorithms they employ. By shifting conservatism factors or gradient factors you can match different computers fairly closely.

REBREATHERS AND COMPUTER ALGORITHMS[9]

Manufacturer	Model	Computer	Bühlmann (DS)	VGM	GF	VPM-B	DCAP	FUSED RGBM
AP	Inspiration Evolution	Vision			x			
Innerspace	Megalodon	Shearwater			x	x		
rEvo	rEvo III	Shearwater			x	x		
Poseidon	MKVI Discovery	Poseidon					x	
VR Technology	Sentinel	VRX	x	x				
JJ CCR	JJ	Predator			x	x		
Hollis	PRISM2 Explorer	Shearwater Special DG05	x	x	x	x		

CHAPTER 7 - PROCEDURES

Manufacturer	Model	Computer	Bühlmann (DS)	VGM	GF	VPM-B	DCAP	FUSED RGBM
Liquivision	computer only	X1			x	x		
HW	computer only	OSTC			x	Possible		
Shearwater	computer only	Predator Petrel			x x	x x		
Suunto	computer only	DX						x

ALGORITHM COMPARISON CHART[9]

The chart on the next page compares dive times generated by common algorithms in different rebreathers and dive computers. The comparisons are made to depths up to 45m with air as diluent. This keeps the comparisons within the scope of recreational rebreather diving. The following data have been obtained using real data from the Suunto DX at the 5 personal adjustment settings plus the use of the following planning software:

- Gradient Factors- AP Projection
- VPM-B- V-Planner
- DCAP- Poseidon WeDive
- VGM- VR Proplanner

Comparisons were made with some commonly used settings as described below:

- Gradient Factors as recommended from the AP Inspiration Vision manual 90/95, 50/90 and 15/85
- VPM-B settings of 0 and +2
- DCAP in Poseidon WeDive app replicates the Poseidon MKVI actual dive data
- VGM carried out with zero bubble control

As far as possible all other settings were kept the same (descent and ascent rates, etc.) The following data are intended as a guide only and should be individually verified.

Ken Smith assists John Dalla-Zuanna with his Flex sidemount rebreather as he prepares to return to the water for his second dive of the day.

CHAPTER 7 - PROCEDURES

Air Diluent	Setting									
Algorithm	Depth Time	30m :60	33m :60	36m :60	39m :60	42m :60	45m :60	30m :45	35m :45	40m :45	45m :45
Suunto Fused RGBM	P -2	74	82	92	106	113	118	52	62	72	82
Suunto Fused RGBM	P -1	76	86	97	111	119	124	54	64	74	86
Suunto Fused RGBM	P0	78	88	99	116	124	129	56	65	76	88
Suunto Fused RGBM	P1	82	96	109	124	132	137	58	67	80	94
Suunto Fused RGBM	P2	87	102	116	131	138	143	59	72	86	99
Gradient Factor	90/95	75	82	90	97	106	116	53	63	72	83
Gradient Factor	50/90	76	84	92	100	110	120	55	64	74	86
Gradient Factor	15/85	78	86	95	105	116	127	56	66	78	90
DCAP	none	79	86	95	104	117	127	54	66	77	90
VGM	0	78	86	97	108	121	135	54	63	75	88
VPM-B	0	71	77	85	92	101	110	52	60	70	81
VPM-B	+2	73	80	88	97	106	115	53	63	73	85

Repetitive Dives

If the remaining duration of your consumables allows, it is completely normal to dive again without filing tanks or replacing your scrubber. When you get out of the water, secure your rebreather according to manufacturer's suggestions. In some cases this might even include powering down the unit to conserve batteries or closing cylinder valves. On a noisy boat, you might be advised to turn off your tanks so you don't miss out on hearing a leak. Some rebreathers have protocols for draining some of the metabolic moisture that has accumulated in the breathing loop. For this reason, it is absolutely critical that you complete a pre-dive check and pre-breathe prior to entering the water again. Many accidents have occurred because a diver failed to turn a tank back on or failed to seal the loop properly. Your rebreather may come with an abbreviated checklist for use between dives. The onboard computer will keep track of your profile from the first dive, but you might need decompression software on hand to plan the second dive appropriately. Some handsets have a simulator/planner feature that helps

CHAPTER 7 - PROCEDURES

with quick dive forecasting. If this is not the case, I suggest getting deco planning software for your smartphone or tablet so you can have it close at hand on a boat or at a dive site. I have automated my checklists on my iPad. I have the iPad available not only for checklists, but also in the event I need to access the instruction manual or user service manual for my rebreather.

Packing the Canister

One important step in your pre-dive preparation includes packing the canister. Recreational rebreathers are sometimes designed for pre-packed canisters. These canisters might be

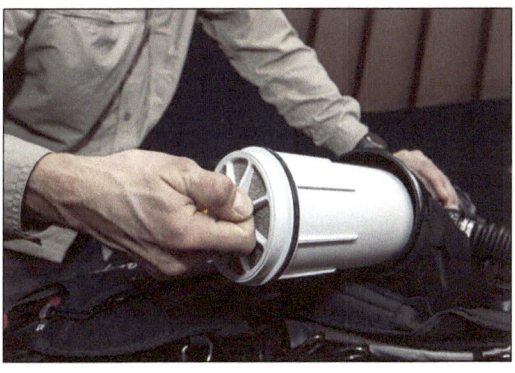

Scrubbers are sometimes supplied pre-packed and others are packed by the user. In either case, it is critically important that the CO_2 seal or O-ring(s) are inspected and properly installed. Absence of this critical seal can lead to rapid carbon dioxide buildup, unconsciousness and death.

the ExtendAir type or pre-packed granular absorbent. If your rig uses a pre-packed canister, you will simply need to remove the packed scrubber from its shipping package and insert it into your rebreather following the manufacturer's instructions while ensuring the CO_2 seal and/or O-rings are properly installed.

Many technical rebreathers are fully prepared by the user when their scrubbers are filled with granular sorb. Proper packing takes a little patience. Strict adherence to the manufacturer's guidelines is critical. Generally, the sorb is poured into the canister in stages. After each stage, some form of tapping or shaking helps to settle the contents. Each canister has its own directions including steps to pour, shake, tap, settle, clean and properly seal. The last step includes final inspection to assure that the contents are packed sufficiently to prevent further settling during transport.

During canister packing, it is important to protect your eyes, hands and airway from crushed absorbent dust. Your instructions may include a recommendation to wear gloves, eye protection and a dust mask.

There are two common and potentially deadly errors in scrubber preparation. The first occurs when not enough patience is taken in properly packing the material. Channeling occurs when sorb is packed unevenly and the gas is able to travel through a path of less resistance, through a gap in loose material. The gas may not have enough dwell time in the sorb and can bypass carbon dioxide through the canister. The second problem occurs when the CO_2 seal/O-ring(s) have not been installed or checked, thus allowing carbon dioxide to pass through without being scrubbed. The best prevention is to carefully follow manufacturer's guidelines when packing a canister.

CHAPTER 7 - PROCEDURES

Pre-dive Checks

Each rebreather has its own protocol for proper preparation. This should include some sort of checklist(s) that have been generated by the manufacturer. They may include:

- **Inspection Checklist**- used for inspecting individual parts prior to assembly. This is convenient when you have not used your unit in a while since it helps you reacquaint yourself with parts and subassemblies.
- **Assembly Checklist**- used for putting the rebreather together. Some manufacturers provide detailed and abbreviated versions so you can use the appropriate one based on your experience. Some of the best detailed lists are fully illustrated.
- **Prep/Functionality Checklist**- used when you are preparing for a dive. This might be conducted the night before or on the morning of a dive to ensure everything is working well and batteries are at an acceptable voltage.

Dr. Kenny Broad uses a checklist to prepare his Meg rebreather for a dive in Abaco.

- **Immediate Pre-dive Checklist**- used within one hour prior to diving or as recommended by the manufacturer. This ensures that everything is fully functional and turned on prior to entering the water.
- **In-water Checklist**- a waterproof reminder card of last minute checks prior to descent.
- **Post-dive Checklist**- used after a day of diving to remind the diver about proper cleaning procedures and recording of consumables.
- **Repair/Maintenance Log**- used to track any anomalies, repairs or routine maintenance on the rebreather.

Some rebreathers are equipped with fully automated pre-dive checks. When the unit is activated, the rebreather computer system walks the diver through a step-by-step functionality test. The best automated systems do not allow the diver to skip steps without successful performance by the unit. In other words, if you don't turn on a tank when prompted, the unit will not allow you to proceed to the next screen. Even though your rebreather has an automated checklist, you will still keep some form of records on repairs, anomalies, sensor voltages, sensor changes, weight requirements, dive logs and other items.

CHAPTER 7 - PROCEDURES

Button Pushers

The Cis-Lunar MK5P rebreather used automation to walk the diver through each step of the unit preparation including the pre-breathe sequence and confirmation of the diver's identity. Some divers quickly realized that they could skip most of these repetitive screens by simply pressing the right button something like 27 times. I once observed a diver skip pre-dive steps in this way. He stood on the boat deck breathing his loop while donning his fins and jumped into the water. Moments later, he was unconscious on the surface, having failed to engage the oxygen injection system. Luckily he was rescued successfully. Automated checks are only useful if they are used as intended.

Calibration Issues

One very important step in your rebreather preparation includes oxygen sensor calibration. Calibration technique is unique to every model of rebreather. Some units must be calibrated every day or every dive. Yet others are calibrated less often and "hold" their calibration longer. Some units are calibrated in two steps with air and then oxygen. Some are calibrated with only a single gas. It's one of the most important actions you can take to guarantee that you are referencing an accurate PO_2 reading.

When you calibrate your rig, you are essentially teaching the unit to read PO_2 correctly. The manufacturer has applied mathematical algorithms for dealing with different temperatures and humidity, but it is ideal if you can calibrate near operational temperature or when the rebreather is warm. Although the specific calibration technique asks a diver to apply a specific fraction of oxygen for a given duration, you are referencing the known partial pressure of that gas, not the fraction of that gas. In other words, when your checklist tells you to flood the loop

In some cases, calibration is done with the unit fully assembled using onboard oxygen. In other cases, a head-only calibration kit is used to conserve gas. Still other rebreathers are calibrated using only air.

CHAPTER 7 - PROCEDURES

with oxygen, it is essentially telling you to apply a known partial pressure of 1.0 to the loop. If the loop is over-pressurized, you may be inadvertently applying a higher reference PO_2 and therefore resulting in a higher partial pressure.

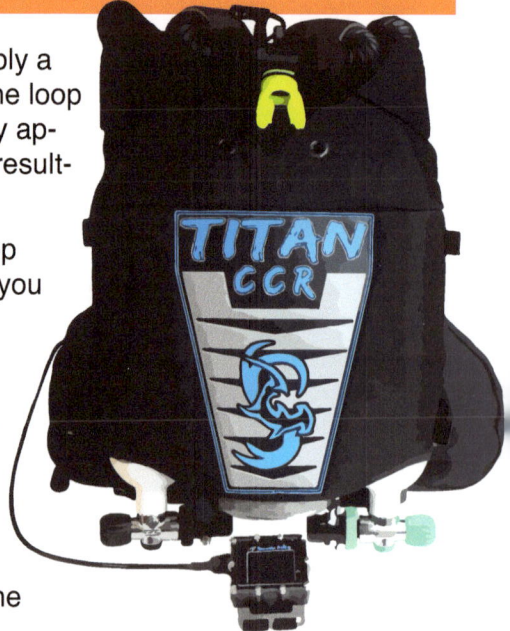

How does this happen? If you flood the loop with pure oxygen and the loop is closed or if you utilize a "head-only" calibration kit that traps oxygen rather than flowing through, then the loop or head can become over-pressurized. You may be creating a partial pressure of 1.1 ATM instead of 1.0 ATM. If the diver confirms their oxygen calibration in this state of over-pressurization, then all subsequent readings will be inaccurate. The result of the above action is a reading that is lower than the actual PO_2 in the loop.

Calibration takes time, patience and strict adherence to the manufacturer's suggested technique. If you doubt the results of your calibration, recalibrate for accuracy.

Altitude Calibrations

If you travel to a region where you will be diving at altitude, you may need to find out the approximate altitude above sea level for accurate calibrations. If you fail to recognize the part that altitude plays in your calibration, you risk giving your onboard computer inaccurate pressure data that will result in an inaccurate calibration. Some rebreathers do this automatically and others require you to input a conversion factor or altitude.

Oxygen sensors read PO_2 (partial pressure), not FO_2 (percentage of oxygen). As the altitude increases, the pressure drops, therefore the partial pressure does too. At 10,000 feet above sea level, your ambient pressure is 0.72 and the PO_2 of air will be 0.15.

For example: If you utilize a head-only calibration kit with pure oxygen, you will flood the head with a pressure that is not equal to 1.0. The higher the altitude, the lower the pressure inside the head. If you calibrate without letting the computer know that you are at a lower atmospheric pressure, then you will be informing the computer that a lower pressure of, let's say, 0.80 is equal to 1.0. At depth you may think you are breathing a 1.2, but are only diving a 1.0. This incorrect calibration could result in hypoxia or decompression illness.

If your unit does not support calibration at altitude, you will need to calibrate your unit prior to departing for altitude. If you must stay at altitude for a series of diving days, you will not be able to re-calibrate with accuracy and may need to

CHAPTER 7 - PROCEDURES

consider using a different rig or diving open circuit instead. Otherwise, cell replacement, barometric pressure changes, humidity variations and other factors could prevent you from using your rebreather.

Preventing Calibration Issues

The most common way that people register inaccurate calibrations is by failing to flush adequately with pure oxygen. This can occur if you either fail to flush enough gas through the loop or accidentally exhale into the loop in the process of purging the breathing loop.

The easiest way to avoid calibration issues is to review this checklist:

✓ Analyze oxygen for purity and accuracy.

✓ Know your altitude.

✓ Ensure a thorough flush of the loop or head prior to registering calibration (Note: loop can be as large as 5 liters and may require 3 evacuations for a full flush).

✓ Refer to millivolt readings as reference (if your unit allows).

✓ Ensure that the loop or head is not over-pressurized.

The head calibration kit for the Optima rebreather.

✓ Replace oxygen sensors early and often.

Voltage Checks, Linearity and Current Limitation

Calibration at sea level produces an output from the cells of 1 ATM or 1 bar PO_2. After calibrating your rebreather in pure oxygen, your cells should return to 0.21 when exposed to air. If they do not, you should re-do the entire calibration and check to see if the cells recover properly. If they don't, you should replace the offending cell(s).

Many rebreathers allow the diver to read the voltage of their sensors through their computer display. If your rebreather has this feature, you should keep a good record of millivolts in air and oxygen. The voltage should be linear, meaning that if your sensor reads 10.0 mV in air, it should read 47.6 mV in pure oxygen. Prior to confirming your oxygen calibration, you should ensure that your sensors are indeed linear and reading correctly.

At sea level, take the mV readings in air and multiply by 4.76 to come up with the predicted mV in pure oxygen.

If you are trying to calibrate with a percentage of oxygen that is less than 100% pure, then you should take the percentage and divide by .21 and use that multiplier instead of 4.76.

CHAPTER 7 - PROCEDURES

For example, if your rig allows you to calibrate with 80% and that is what you are using, then you would take .80/.21 = 3.81. If your sensor is 10 mV in air, then your mV reading in 80% should read 38.1 mV when flushed with EAN80.

Diving at a setpoint greater that 1.0 may leave you vulnerable to hyperoxia unless you can verify the cells are able to reach higher PO_2 under greater pressures. A simple trick for checking if the current hits a limit is to do a robust oxygen flush at 20 feet/6 meters while watching the cell displays to ensure they not current limited. When flushing with pure oxygen at this depth, you should read 1.6 or very close to that on your displays.

If your cells are old, they may not be able to read high PO_2 levels properly, despite the fact they may be in range at lower partial pressures. At 20 feet/6 meters, with a proper oxygen flush, you know what is in the loop. If the cells read low, then you know your sensors are bad.

Even if you are only filling with air and oxygen, you must always analyze your gas. Calibration is only effective if you know the exact contents of your gas.

Completing this drill at the beginning of a dive could be problematic unless you take time to flush back down with diluent. If you are about to descend you need to dump all that fresh oxygen out of the loop so your PO_2 does not spike on descent. The process can use a significant amount of gas. I usually reserve this drill for the end of my dives to check to see that the cells are still behaving properly. I also do this drill at the beginning of a dive if my rig hasn't been in the water in the last week or two. On a re-acclimation dive, the high use of oxygen at the beginning of a dive won't matter so much. One these dives I am usually puttering around in shallow water practicing drills and checking to see that the unit is working properly too.

Let's look at the implications of continuing to allow current limited sensors to control the operation of your solenoid valve in an eCCR. If you have chosen a setpoint of 1.2 and the cells are unable to read that high, then the solenoid may continue to fire over and over while thinking you are below setpoint. If you ever hear your solenoid firing at a very high rate, you might question the validity of your readings and wonder whether your cells could be current limited. Cells that pass a pre-dive check can fail during a dive. Numerous incidents have been reported by divers whose cells performed well during pre-dive and calibration and failed or "locked-up" during the dive. Oxygen sensors are like batteries. You've probably used a flashlight with dying batteries. After a rest period, with the flashlight off, it might fire up properly for a short period of time but soon dims and

CHAPTER 7 - PROCEDURES

goes out. A cell can behave in a similar manner, passing a pre-dive and behaving well at surface pressures, then failing during the dive at higher partial pressure.

Current limitation issues also affect divers whose use mCCRs. In the case above, a diver might repeatedly add oxygen to bring the PO_2 up, not realizing that the cells are limited.

If you want to confirm whether cells are limited prior to hitting the water, manufacturers such as Narked at 90, have created Cell Checkers and Mini Cell Checkers that can give you peace of mind before a diving trip. (http://narkedat90.com). The only snag with these devices is that you will be checking the cells in an environment that is dissimilar to the diving environment and only for a short duration. A cell checker is a mini pressure pot. It is a dry, controlled environment. You'll be exposing your cells to a very moist and warm environment when diving. Studies regarding the value and accuracy of cell checkers are still being evaluated at this time.

Wet Sensors

If you take part in lengthy dives, you may notice that one or more of your sensors will either start to lag in reaction time or read low. This is often caused by moisture that is deposited on the sensor face. This moisture often arises from condensation inside the unit. You should never touch the sensor membrane or attempt to dry it in any way that exposes the sensor membrane to contact with a foreign object or extreme pressure.

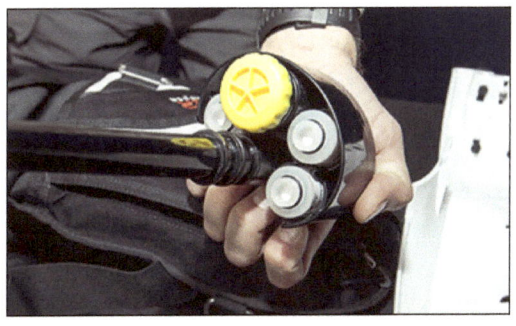

Sensor faces are covered with a hydrophobic membrane. Do not touch the membrane with your fingers or tools.

At times, you may notice a sensor reading that gets "voted out" during a lengthy dive. The "brain" of your rebreather watches the readings on the three independent cells. It uses the average of the three cells to execute commands such as firing the solenoid valve to inject oxygen. If one cell falls out of range of the other two, the computer will isolate and ignore that reading in its average calculations. In this case, the solenoid will fire based on the average of the two remaining sensors, presuming they are functioning properly. This system works most of the time, but not always, since it is possible for two cells to fail at the

CHAPTER 7 - PROCEDURES

same time. Some voting logic includes additional failsafes to account for known issues with cells.

Older sensors that are beyond their useable life may read low in the higher ranges of PO_2 or may completely limit out, even if they calibrate correctly prior to diving and seem reliable at the surface. In rare cases, moisture can get on the back side/electronic contact side of the sensor. In this case, the sensor might read artificially high.

If a sensor is ever "voted out" during a dive or if you suspect a sensor problem you should abort your dive. For technical divers, a diluent flush will advise you which sensor(s) are still reliable. Then you can make a choice about whether to: fly the unit manually on one reliable sensor, allow the voting logic to operate the solenoid based on info from two sensors or abort to SCR or open circuit.

How Old are Your Sensors?

Many sensors are numbered with a date code by the manufacturer. This date code indicates the "born on" date. The codes are usually letter-number combinations. For some brands, if your sensor reads A8, it was manufactured in January 2008. If the code reads D9, then it was manufactured in April 2009. When you remove a sensor from its bag to install in your rig, you should label the sensor to indicate the date you installed it in the unit. Some sensors are now labelled with "discard by" dates. A sensor still ages inside its original packaging. You may have to discard unused, but expired sensors. Check your cell or rebreather manufacturer's website to verify how to read the date codes on your brand of sensors.

Paul Heinerth completes his pre-breathe on his Optima CCR.

Pre-breathe

Your final checklist will include a notation to pre-breathe the unit. This is generally the last step on your checklist. Contrary to popular belief, this action is not intended to "warm up the canister and get it working." If you put a carbon dioxide meter on the outbound side of a rebreather canister, it will show that a cold canister begins working immediately. Since this is the case, many people wonder why we pre-breathe at all. If you have ever owned a rebreather with a device that monitors temperature of the sorb bed, you will observe how the absorbent warms as you breathe on the canister. It takes about five minutes to move up to a high temperature. During

CHAPTER 7 - PROCEDURES

this time, you have the opportunity to discover other faults in your preparation such as absorbent channeling, absence of canister, CO_2 seal functionality, solenoid operation, gas injection, proper functionality of mushroom valves, etc.

The pre-breathe should be conducted in a safe environment while you pay close attention to your rig. Twice in my rebreather diving career, I have rescued individuals that carelessly walked on the dock and boat

Don't breathe on your loop until you are ready to pay full attention to CCR operation.

while pre-breathing their unit. For various reasons, both passed out and one nearly drowned when he lost consciousness after making a giant stride into the water. Pre-breathes are meant to help us catch those inevitable human errors that we all make on a bad day. Your life is worth a five-minute investment, every time you dive.

The other critical point about pre-breathing is that your nose must be blocked. This can be achieved by either pinching your nose or wearing your mask. Your brain is insidious. If you are having a CO_2 issue and your nose is not blocked, then you will "sneak" air in through the nose without even realizing it. If your nose is blocked it will be more obvious that you are having a problem.

The Danger Zone

One of the greatest danger zones for rebreather divers can be the boat deck or stairs leading to the water (see above!). With fins in hands and a mask on the forehead, it is not unusual for someone to slip and fall into the water unprepared. Several incidents have occurred on lurching boats and slippery stairs. There is no reason to walk while breathing from your loop. You should be paying attention to your feet. The loop constricts your field of vision and you are not likely to properly monitor displays while walking.

As you approach the danger zone, ask yourself and ensure the following:

✓ Tanks on?

✓ ADV enabled?

✓ Safe PO_2 in the loop?

SAFETY CHECK

Do not pre-breathe your CCR while walking. Sit down and pay attention in a safe location in case you pass out!

In the event you fall in the water, you will be able to breathe confidently from the loop while you shake off the urge to feel embarrassed.

CHAPTER 7 - PROCEDURES

Fire!

A scientific diving colleague of mine recently shared a story that could have ended very badly. He was borrowing a rebreather from a university dive locker. He thought the unit looked a little rough, but it passed all the pre-dive checks with no issues. He sat on the transom of the boat and began his last minute preparations, fully dressed and ready to dive. He reached back and turned on his tank valves. He heard a small pop and then mere moments later was hit like a football receiver who had just run full speed into a linebacker. He was vaulted into the water in a blur of confusion. He surfaced angry from his back roll to meet the pale and terrified faces of his crew. He had experienced an oxygen fire without even knowing it. As he turned on the valve, his fellow divers saw a rooster tail of flames erupt from his backside. They acted quickly, pushing him into the water to extinguish the flames. His biggest takeaway lesson was that he will never open an oxygen valve while his rebreather is strapped to his body. His last minute checks will be done before donning the unit.

In-water Procedures

Loop Protocol

If you are on the surface of the water and the loop is out of your mouth, the loop should be in the closed position so you don't accidentally flood the unit. Even if you are taking the loop out for a quick word with your diving partner, you have to get in the habit of closing the loop at all times, with no exceptions.

Check your displays immediately before breathing from the loop. As you check the actual PO$_2$, ensure that a low setpoint has been enabled. This setpoint will likely be between 0.5 and 0.7. If a setpoint is not enabled, you might breathe the loop down to hypoxic levels if you are not paying attention. Some rebreathers have warning systems to prevent this, while others do not. Yet, even with the most advanced warning systems, I have witnessed distracted divers breathe a loop to hypoxia because they failed to turn something on.

After ensuring safe PO$_2$ and setpoint, put the DSV in your mouth, blast the DSV clear of water and then open the loop. (This will help to keep the loop clear of excess moisture.) Recheck displays once you are breathing.

CHAPTER 7 - PROCEDURES

In-water Checks

Some agencies and manufacturers have created safety cards or slates describing the in-water buddy check. Use the one for your rebreather as a last minute check.

✓ Check your BOV or offboard second stage. Breathe from it.

✓ Give your buddy a head to toe safety check of all critical safety gear for the environment you are diving in.

✓ Check displays.

✓ Check that your eCCR is injecting appropriately one last time.

✓ Ensure safe PO_2 and setpoint and start your descent.

Christian Clark and Phil Short check each other for leaks before continuing their dive.

Descent and Verification of Safe Operation

Once you are free of the effects of surge, pause your descent to complete a bubble check and ensure everything is operating properly.

Descending on a rebreather may feel a little different than open circuit diving. There is lot more going on and you are managing an additional airspace contained within the breathing loop. If you are weighted properly, it might take a little extra effort to get off the surface and begin your descent. You should barely float on the surface with a minimal loop volume and empty BCD and dry suit. To descend, dump any air from your wing and your dry suit if applicable. You will also need to exhale through your nose to lower the loop volume. Once your head is underwater everything gets a little easier. You will almost completely expend the gas in the counterlungs and feel a bit of a brief vacuum. The ADV should fire, injecting diluent into the loop. If it does not, your rebreather may be equipped with a manual diluent injector that you can use, but most commonly, the problem is a fit issue. If the rebreather fits properly, the ADV should inject diluent on its own.

CHAPTER 7 - PROCEDURES

Breathe normally as you descend. Watching your displays, you will see the PO_2 rise from the increased pressure.

Somewhere around 15 feet/4.5 meters or where convenient, pause in your descent to conduct a bubble check with your dive partner. Get in a horizontal, neutrally buoyant position and check your buddy's unit for any escaping gas that is not part of normal operation. Look for any equipment that is not properly fitted. Have your buddy thoroughly check your unit. Check all systems and displays of your own to ensure that everything is fully operating. If anything fails this test, abort the dive. No leak should be ignored. To complete your underwater safety check, breathe from your BOV and bailout then ensure all hoses are stowed properly. Conduct a current limitation check of sensors if applicable. Now you are ready to continue your descent.

On a normal descent you may not hear the solenoid fire at all and if you are operating a mCCR, you may not need to inject oxygen. Once you have reached depth your rebreather should be switched to its high setpoint either automatically or manually. Some rebreathers will conduct automated self-checks in the first few minutes underwater. You may be able to listen for those or watch your display for the series of operational checks and calibrations.

Bottom Time

During your bottom time, you need to remain vigilant about checking your displays. There are a variety of parameters that may be monitored on your primary display, HUD and secondary display if you have one. You must always know your PO_2 and cross-check it periodically. Approximately every one to four minutes, you should actively check displays.

Common display features include:

✓ Actual PO_2 average

✓ Actual PO_2 on each sensor

✓ Setpoint

✓ Alarm status

✓ Depth

✓ Time

Less common display features:

✓ Actual CO_2 (very few rebreathers)

✓ Remaining scrubber time

✓ Scrubber status (thermal profile)

✓ No stop time remaining

✓ Decompression status

Throughout your bottom time you will be frequently checking displays.

CHAPTER 7 - PROCEDURES

- ✓ Time to surface
- ✓ Remaining oxygen pressure
- ✓ Remaining diluent pressure
- ✓ Battery status and alarms
- ✓ Diluent mix
- ✓ Bailout mix
- ✓ Bailout gas pressure
- ✓ Ascent speed
- ✓ Time remaining (based on all dive parameters)
- ✓ Bailout tank pressure(s)

Peter Sotis helps a student build on his understanding of the computer interface, revealing new features sequentially.

Your rebreather may require you to manually shift your setpoint to the high setpoint once you reach the bottom. Other rebreathers do this automatically, by floating the setpoint upward until it reaches the high preset you chose before the dive.

Minimum Loop Volume

Once you have reached max depth, you will adjust the volume in your loop to achieve "Minimum Loop Volume." In essence, you should be able to take a full breath comfortably. At the end of the inhalation of a full breath, you should begin to trigger the ADV. Diving with this volume in the loop will make buoyancy control easy. If you are diving at a stable depth, you will slowly metabolize molecules of oxygen which in turn will lower the volume of gas in the loop. As the volume lowers, so does the PO_2. If you are diving an eCCR, this will trigger the solenoid to fire just enough to bring you back to setpoint without dramatically changing volume and causing buoyancy shifts. If you are diving an mCCR you'll inject the gas with the manual injector in small bursts. In addition to your displays, the slight drop in buoyancy from oxygen metabolism or the slight loss of volume in the breathing loop will let you know it is time to inject a small amount of oxygen.

When you master minimum loop volume, buoyancy control will become intuitive.

When you are trying to find that sweet spot of Minimum Loop Volume, you may have to vent a little gas around your mouth or through your nose. Try taking another breath to see if you can pull the ADV at the end of a deep inhalation that is slightly greater than your normal breath volume. If not, vent a little more until it feels right.

CHAPTER 7 - PROCEDURES

If you dive with a greater loop volume, you may find that you are routinely triggering the OPV and losing precious gas. This is all tied to proper weighting as well. If you vent repeatedly, you will sink and trigger the ADV, thus diluting the gas in the loop. When that happens, your solenoid will fire to bring up the PO_2 and it creates a vicious cycle of venting, descending, diluting and adding rich oxygen, then rising, filling, dumping, etc. It is completely normal to struggle a little with this until you get the "feel" of a rebreather and can anticipate the pace of the solenoid and optimize your weight and OPV setting. Be patient, this is the one thing that frustrates an experienced open circuit diver until they get the hang of it.

Maintaining Buoyancy

Buoyancy control is significantly different than open circuit SCUBA. If you are an experienced diver, then you have mastered the careful breathing that controls your buoyancy. Now that you are breathing in and out of a flexible enclosed space, you buoyancy will stay constant throughout the breath cycle. If you maintain a relatively constant loop volume, then buoyancy will remain constant. Buoyancy is controlled with proper weighting and air added to the wing or drysuit.

Given that your onboard tanks are very small and the gas used is very limited, you will find that your buoyancy remains stable throughout your dive. In open circuit diving you will actually use up several pounds of gas during a dive and will become more positively buoyant towards the end of the dive. This is not the case with a CCR. If you are weighted correctly at the beginning of a dive, you should stay that way. The only thing that will make you more positively buoyant is an open circuit abort scenario where you use a lot of gas from your offboard tanks.

Kevin Gurr and Richie Kohler (behind) demonstrate that hovering is easiest in good horizontal trim.

You will find that you dive a little differently for optimal buoyancy control and easy CCR operation. If you are able to stay at a constant depth and swim around something, it may be preferable to swimming through a lot of "ups and downs." Each micro ascent may cause you to vent bubbles and each micro descent may

CHAPTER 7 - PROCEDURES

cause the ADV to fire and solenoid to catch up and readjust the loop PO_2. This uses more gas and takes a little more effort than simply swimming around an obstruction.

Ascents

When it is time to ascend, you and your buddy signal your readiness. As with open circuit, ascents are slow and controlled. As soon as you begin swimming up you will notice the loop gas expanding. Depending on the rebreather model, you will either enable the OPV to vent or will dribble excess gas out around your mouth and through your nose. A combination of both techniques may also be used. You will expect to hear the solenoid firing quite a lot in an eCCR as it tries to catch up with the naturally dropping PO_2. Most eSCRs will lower the float-

Kim Smith and Curt Bowen use an ascent line to return to the boat.

ing setpoint on ascent to minimize the bubbling. On an mCCR you should work hard to anticipate changes. Vent gas proactively, add oxygen and circulate the loop in a way that conserves gas and minimizes needless venting. Even eCCR operators may use some manual control on their ascent to anticipate changes and minimize venting rich gas.

As you swim up, watch your displays closely and continue a slow ascent rate. Remember to vent gas from your wing and dry suit if applicable. If for some reason you feel yourself losing control of your ascent, vent breaths, flare your legs out or give yourself a bear hug (for OTS counterlungs only) to reduce the volume in the counterlungs. Then recheck your displays and restore the proper PO_2.

When you reach a safety or decompression stop, you will find it easier to float horizontally in the water column, because as soon as you shift to a vertical position, the pressure differential may cause venting to occur. Your ascent may include using a DSMB if you are not using an anchor line or mooring.

Reaching the Surface

When you reach the surface, establish positive buoyancy, close the loop and remove it from your mouth. If you forget to close the loop, flooding can cause rapid negative buoyancy. Take

CHAPTER 7 - PROCEDURES

a moment to stow your DSMB so you do not get entangled when you try to get on the boat. Good surface habits are important since accidents can happen when divers mistakenly think that all dangers have passed.

Rebreather divers don't often wear a standard snorkel since it gets in the way of the loop, but a pocket snorkel can get you back to the boat easily. If you must swim on the surface using your rebreather, be sure to carefully monitor displays and actively inject oxygen if you are using an mCCR. Numerous incidents have occurred when divers failed to monitor displays on a surface swim.

You may remove your bailout bottle prior to exiting the water. If it is supplying gas to your wing or dry suit, you should secure positive buoyancy first, then undo the LP inflator and pass the tank up to surface personnel. When you climb a boat ladder or conduct a shore exit, it is a good idea to breathe from either the loop BOV (as long as your diluent is not hypoxic) or a bailout bottle in case you fall back in the water. Maintain normal operation and monitoring if you are on the loop and understand that if the loop is dislodged in a fall, it can flood and cause negative buoyancy. Always know where you can get the next safe breath. If you already passed up your bailout tank, then perhaps you can breathe open circuit through the BOV judiciously as you climb the ladder.

Once safely onshore or the vessel, remove your unit and follow manufacturer's instructions for safe shutdown. In some cases, you may not need to do anything, but if you are using an SCR with constant flow, then you will need to close tank valves to stop the flow of gas. If you plan to do a followup dive, prepare yourself for a pre-dive check and pre-breathe before re-entering the water.

No Excuses for Poor Trim

Everyone gains a little bulk when diving a rebreather, but it is no excuse for poor trim. Stage bottles can be carried cleanly, but the technique for carrying them varies from traditional open circuit SCUBA gear.

With over-shoulder counterlungs, the chest D-ring can be difficult to reach. Even if you are able to clip in at that location, the tank may interfere with ADV operation, dump valves, etc. A technique borrowed from sidemount diving helps you place the bottle into a better streamlined path. It makes the bailout regulator easier to find and offers you solid streamlined trim.

The bottom of the bottle can be clipped into a "butt-plate" manufactured by Dive Rite, Golem Gear or others. Some manufacturers carry a double

CHAPTER 7 - PROCEDURES

D-ring (butterfly shaped) that slides up the crotch strap and can be used in a similar fashion to a butt plate. The tank valve is captured by a bungee cord instead of being clipped to the chest. The bungee attaches to the backplate at about the bottom edge of the shoulder blade and the top end of the bungee is clipped to the harness swivel or chest D-ring. The clips on the stage bottle should be installed roughly opposite from the tank valve hand-wheel so that the bungee will slide easily over the hand-wheel. When the hardware is mounted in the correct orientation the hanging valve will not fall out of the bungee cord.

When the bottle is properly hung, it will ride lower than a traditional stage bottle with the valve tucking easily under the armpit of the diver. The tank should run parallel to the diver's side in good trim. Tanks are quicker to remove, replace and regulators are easier to access. Swimming trim is improved and the environment is protected from unnecessary damage.

Mixed Teams

When rebreather divers and open circuit divers dive together, they are referred to as a Mixed Team. OC divers are often too embarrassed or intimidated by the advanced technology to ask a rebreather diver about differing procedures. It is incumbent on you to ensure that gas management and emergency procedures are clear prior to entering the water.

When open circuit and closed circuit parters dive together, they are referred to as a mixed team. Diving with someone who is using a CCR that you are not familiar with also creates a need for orientation.

Orientation

✓ Show how the rebreather is worn and how it can be removed.

✓ Demonstrate how the wing is inflated and, if it is attached to an onboard cylinder, discuss how this limited supply could be easily exhausted during a rescue. Determine whether oral inflation of the wing by the buddy is possible.

✓ Discuss how developing problems can be quickly recognized.

✓ Demonstrate various warning lights, especially those that indicate life-threatening oxygen levels.

✓ Describe the significance of a vibrating mouthpiece, if applicable.

✓ Describe how and when it might be necessary to close the loop, and why preventing a loop flood is critical for maintaining buoyancy.

✓ Show how to operate and purge the BOV.

CHAPTER 7 - PROCEDURES

✓ Practice sharing gas. Determine whether sharing a long hose or passing off a stage bottle will work better.

Gas Planning

Inquire about the SAC Rate of the OC diver and plan appropriate gas volume to ensure their safe ascent using your open circuit bailout gas.

✓ Select bailout gas that is compatible with the OC diver's decompression plans.

✓ Plan decompression gases to accommodate all emergency scenarios for either diver.

✓ Discuss how the team will stay together when you reach decompression stops. The CCR diver will likely complete their deco much earlier than the OC diver. Are they willing to wait for their buddy in the water?

Complete a Safety Drill

Describe what to look for during the bubble check. Rehearse gas-sharing scenarios before descending so the OC diver is certain about where to find a second stage.

If rebreather divers strive to maintain a high level of conservatism and independence with their bailout gas, then safety and flexibility are benefited. Self-rescue is assured and buddy-rescue of a CCR or OC diver is also probable. The goal of the orientation is not to teach the OC diver how to run a rebreather, only how to share gas and handle emergencies that may occur.

You should check your BOV and bailout second stage underwater to ensure everything is fully operational and to improve your motor skills and response.

Gas Management

Catastrophic failures on open circuit SCUBA are usually manifested in events like high-pressure seat failure in a first-stage, hose rupture or manifold damage, burst disk and valve breakage. Technical divers spend ample time rehearsing valve drills and abort scenarios, since gas loss equates to time pressure. They manage the emergency and abort the dive.

On a rebreather, failures are more likely to develop slowly. In some ways, a hose rupture or first-stage failure is one of the easiest issues to deal with. In many cases, the diver simply reaches back, turns off the valve and feathers it on and off through ascent or simply aborts the dive using open circuit gas. Because of the slow onset of issues, OC divers should be instructed to watch for behav-

CHAPTER 7 - PROCEDURES

ioral changes in a rebreather partner that could indicate hypoxia. Rescuers have often reported strange swimming patterns or odd behavior before a partner diver passed out.

In the early days of rebreather training, we used to put a considerable emphasis on keeping the diver on the loop. These days, we teach students the myriad options available to them in emergency scenarios, but encourage divers to bailout to open circuit if they have any doubt about the safety of what they are breathing, or if the task load is too large. *If in doubt, bail out.* For recreational sport rebreathers, bailout is the only option. OC dive partners should understand the importance of monitoring a rebreather diver who is having a problem and they should be prepared to offer assistance if an OC bailout occurs.

Problems Take Time to Develop

I was involved in a serious accident investigation that had a rare happy ending. It is not often that a survivor can be interviewed about an incident that included unconsciousness, near drowning and successful resuscitation. The rebreather was equipped with a black box that we could download to review changing status throughout the profile. The rebreather had passed its pre-dive checks and performed properly in the first few minutes of the re-acclimation dive that had the victim practicing drills while monitoring a junior diver. Several minutes into the dive two sensors became current limited and failed simultaneously. The investigation uncovered that the cells were well beyond their expiration date. What is interesting, is that the diver ignored flashing lights and warnings that indicated that he had a cell failure. Assuming that voting logic was taking care of him, he continued his dive. As a result, he was breathing a continually rising PO_2 as the solenoid tried to bring up the oxygen based on two failed cells. He breathed above 1.6 and as high as 3.0 for over 20 minutes before experiencing a seizure underwater. There are numerous lessons to learn from this incident, but what is interesting is that the diver experienced over 20 minutes of flashing lights and serious warnings on his rebreather. Had he acted on them, the dive could have ended with much less drama. Problems often take time to develop on a rebreather. Every warning should be heeded and cross-checked. Alarms should not be suppressed unless you have cross-checked and solved the problem.

---

[7] Basic Cave Diving: A Blueprint for Survival, Published by the NSS Cave Diving Section, Sheck Exley, fifth edition, 1986

[8] Calculation of Air Saturation Decompression Tables, Workman, RD, 1957, US Navy Experimental Diving Unit, Project #NS185-005

[9] Crawford, Ryan. Suunto DX Fused RGBM Comparisons in CCR Mode as presented at the Suunto Global Distributors Meeting, September 2013

Checklists Save Lives

Use a checklist for preparation. Do a full 5-minute pre-breathe. Don't get in the water if something is not working properly!

8

Failure Modes

In this chapter:
- *Trusting Technology*
- *Accident Analysis*
- *Common Problems and Solutions*

Faster than a speeding bullet

If your vehicle's accelerator got stuck, would you know what to do? The news these days is filled with horror stories of people on cell phones speaking with an emergency dispatch operator, panicking as they careen down the highway, out of control. Unfortunately, some of these cases ended in fiery crashes when neither the driver, the passengers nor the operator thought to shift the car's transmission into neutral.

If your solenoid valve on your rebreather stuck open, would you know what to do? If a hose blew on the oxygen tank, could you react in an instinctive way?

Several years ago, the rear tire tread came off my van at 65 miles per hour on the Florida Turnpike. I knew instinctively that using the brakes would send us into an irrecoverable roll. Many years of driving on snowy Canadian roads, gave me the manual practice that still left its imprint after 15 years in America. We skidded down the highway sideways while I continued to try to steer us out of the skid and ease to the shoulder and out of traffic. By the time we impacted the soft shoulder, we were less than half speed. Although we still flipped the van, I think it could have ended much worse. We were virtually uninjured.

Anyone that has dived rebreathers for a long time will share that they have had bad days when either the equipment failed or they made stupid mistakes. If your skills are well practiced and manual responses are instinctual, then your chances of survival will be far greater. Practice doesn't always make perfect, but survival doesn't have to be pretty, just effective.

CHAPTER 8 - FAILURE MODES AND EMERGENCIES

Trusting Technology

CCRs are getting more reliable and feature-rich. Fewer equipment failures have resulted in more trusted acceptance of this type of technical diving. However, I am grateful to have learned in the earliest days- the era of "breakage." For my first years of CCR diving, it was rare to have a dive that went off without a hitch. As a result, I grew vigilant with the notion that my technology could fail me at any moment. That made me conservative with bailout gas and meticulous about preparation and safety. To this day, I still run a full check the night before my dive. I get up in the morning and do it again in the garage. Then, if I am on a boat, I check it prior to leaving the dock and again with my pre-breathe sequence prior to diving. Sounds excessive? It's not really onerous at all, just part of my routine that allows me to jump in the water relaxed and prepared. When you are certain the equipment is fully functional before you hit the water, it leaves a fully engaged brain for anything unexpected.

CCRs are extremely reliable and you may not experience any real issues in class. Don't allow complacency to slip into your daily regimen just because "nothing has happened yet"!

Recently, a friend of mine had a very lucky day. He came home from a very bad dive. In critiquing the incident I asked, "why didn't you bail out?" He replied, "I screwed up. I was so confident with the gear that I knew it wasn't the rebreather." Before he knew it, he was impaired and barely able to make a switch to open circuit. Not recalling the end of his dive, he surfaced on sheer fortune. He was "fuzzy" for hours. An experienced rebreather diver, but one without any incidents in the last many years, he had become over confident and complacent. Be careful out there and retain a healthy amount of fear in your diving.

Accident Analysis

When Kevin Gurr of VR Technology first considered designing rebreathers, he did an exhaustive examination of rebreather incidents and accidents and created a list of common issues that led to problems.

The following list of fifteen points describes his findings on the root cause of the majority of fatalities and serious incidents:

✳ Divers started the dive with their electronic control system off.

✳ Divers started the dive with their oxygen turned off.

CHAPTER 8 - FAILURE MODES AND EMERGENCIES

* Divers descended with diluent off and then panicked when they could not find the manual addition.

* Divers did surface swims on hypoxic diluents.

* Divers did not pack the absorbent canister correctly or the design of the canister allowed CO_2 to bypass if O-rings where incorrectly greased or assembled.

* With insufficient guidance on canister durations, people exceeded the duration limits.

* Temporary floods made the breathing loop unusable.

* Insufficient filtering produced oxygen solenoid failures.

* Hose attachment systems produced stress-points which destroyed the hoses.

* Electronics in the loop were affected by moisture.

* Gas supplies were accidentally switched.

* Failures in the electronics made the unit unusable.

* Divers become stressed at high work rates.

* DCI occurred as a result of the unit's inability to maintain a near constant PO_2.

* Divers did not follow pre-dive procedures.

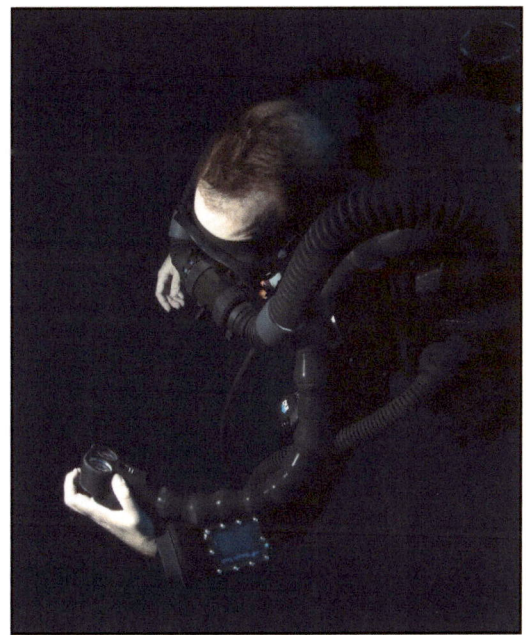

Casey Omholt checking tank pressures on his Nautilus rebreather. Don't overlook your SPGs just because the supply is used very slowly. Gas supplies can diminish quickly in the the presence of a leak.

Design features of his Ouroboros, Sentinel and Explorer rebreathers were geared towards addressing the issues above. But, it is worth considering each point to determine how *your* rebreather would respond to each situation. If a particular rebreather design has a vulnerability, how can you modify pre-dive behaviors and diving practices to prevent the issues Gurr describes? It is easy to say "it can't happen to me," however, statistics prove that well-trained and experienced divers have fallen victim to what appear to be rookie mistakes.

I'll always argue that proper pre-dive checks and pre-breathe sequences will help prevent most accidents, but careful review of the above list in context of your personal rebreather is worth a few minutes of careful consideration.

CHAPTER 8 - FAILURE MODES AND EMERGENCIES

Common Problems and Solutions

The following section describes some common problems and offers solutions. A recreational rebreather is rarely equipped with the features necessary to take these actions, and the resolutions for most of these issues are beyond the scope of recreational training. In the case of any emergency on a recreational rebreather, you should bailout to open circuit unless otherwise instructed in your class.

Confusing Data

When the face of an oxygen sensor gets wet, it may read low and slow. Sensors that get wet on the wiring side can read high, but this seems to be more rare. Aged sensors may also give erroneous readings, especially at higher PO_2 levels. Most eCCRs offer a warning when a sensor drifts out of range from its partners. If you have any doubt about the accuracy of sensor readings, a vigorous flush with diluent gas will help you determine if any of the sensors are reliable and accurate. After determining which, if any sensors, are accurate, you can choose to either allow the system's voting logic to get you home or run the unit manually on a single accurate sensor during your abort. If you are at all uncertain about the accuracy of your sensors, from flooding, poor calibration or other electronic failures, then an open circuit bailout is not just warranted but is the only safe option.

With one oxygen sensor "voted out," Noel Dillon flushes his loop with diluent gas to confirm which of the sensors is accurate. Watching the displays as he flushes, he can see which sensor(s) respond correctly.

When conducting a diluent flush for verification, you should be able to do a little quick math to determine which sensor(s) are good. If you are at 100 feet/30 meters, then the pressure at that depth is 4 ATA. (Check the Physics Chapter if you have forgotten how to turn depths into pressure). If your diluent gas is air and you have done a vigorous flush following the manufacturer's technique, then you should see a PO_2 of 4 ATA x 0.21 = 0.84. If a sensor is "stuck" or reads something completely inaccurate, then you will know it is not safe to trust. It is not uncommon to have two bad sensors. When their age and usage history is identical, it is possible them to fail on the same dive at the same time. Other factors such as unit flooding can cause multiple sensors to fail.

Your computer may offer a screen showing you the expected diluent PO_2 for your depth. Using this feature when you flush with diluent will prevent you from needing to do math underwater. Depending on the type of rebreather you are

CHAPTER 8 - FAILURE MODES AND EMERGENCIES

using, you may have to temporarily shut down the oxygen supply to the solenoid to avoid oxygen injection during this drill. If that is the case, you may have to turn it back on again after the drill. You'll practice this drill several times during a rebreather class.

Diluent Flush

A diluent flush will temporarily solve almost any problem and give you time to think. Given that diluent gas is mixed so that it is always safe to breathe (with some exceptions with hypoxic trimix diving), flushing with diluent buys you time and a known PO_2, even if you don't have electronics.

A diluent flush will help bring PO_2 down if your descent is too fast and your PO_2 starts to climb above desirable levels.

A diluent flush can temporarily solve hypoxic PO_2, hyperoxic PO_2, failed electronics, unreliable sensors and other issues. During your class, you will learn the best technique for conducting a diluent flush. Practice it often and understand how it can buy you time until you figure out what is going wrong on your CCR. Some eSCRs use an automatic diluent flush as their failsafe. When a problem is detected by the system, the diver is instructed to bailout. If they do not heed the warning, the system will revert to a continuous flow of fresh diluent gas to get the diver safely home.

Catastrophic Loop Failure

Mechanical issues may lead to catastrophic loop failures that demand open circuit bailout- ripping or tearing a breathing hose inside a wreck, counterlung tears, dry-rotted rubber hoses, lost or torn mouthpieces and breakage of the DSV mouthpiece lever itself are all examples of failures that render the loop unrecoverable. With water flooding the loop you may become negatively buoyant very quickly. In this case, bail to open circuit, adjust buoyancy and ascend.

Rip and Roll

I was on a dive in the Dominican Republic when my dive partner Curt Bowen looked over at me and yelled, "you, up, now!" We had just reached the bottom and everything felt fine. Operations were normal. He then signaled, "bubbles." I turned my camera on myself and took a picture. Glancing at the LCD screen, I saw a stream of bubbles coming out of the exhalation hose on my Meg CCR. I aborted in closed circuit and ascended safely to the boat. I wondered how such a serious rupture could have been missed during the pre-dive loop pressure checks. As it turned out, the hose bent at close to ninety degrees where it joined the

CHAPTER 8 - FAILURE MODES AND EMERGENCIES

canister. This area had been slowly dry-rotting over time, right at the point where the rubber hose attached to the plastic screw port. I probably broke the hose during my back roll entry. I was likely using more diluent than normal on descent, but did not notice because a reasonable volume was needed to reach our max depth at 140 feet. Very little water had actually entered the loop. The positive pressure from the loop let bubbles escape, though very little water entered. I was able to replace the hose and continue diving later that day.

Dewatering the Loop and Partial Flooding

There are times when your own metabolic moisture will build up in the exhalation hose. You may hear some gurgling as you exhale. There may also be an occasion when the loop is momentarily kicked out of your mouth. You will learn specific techniques for dewatering the loop from a minor flood. There may also be specific techniques presented for dewatering the loop between dives. Since this often requires the diver to open the loop in some way, a full pre-dive check including positive/negative pressure checks is imperative, as always.

Carbon Dioxide Breakthrough

Carbon dioxide issues are the most insidious problems that lead to unrecoverable loop failure. Partial flooding and improper packing can lead to channeling of scrubber material. Failure of the CO_2 seal or using carbon dioxide material beyond its specification may also lead to rapid breakthrough. Damaged non-return valves can also lead to undesirable CO_2 build-up.

During your class, you will be shown strategies for dewatering your loop. In some cases you will remove the loop and switch to a second stage while you shake the moisture down through the hoses.

Slowing your breathing or flushing the loop will not lower the CO_2 in your body. You'll be building up further CO_2 if you do not bail to open circuit. Worse, the buildup of alveolar CO_2 will cause rapid respiration that can deplete your OC reserves quickly and cause panic. For this reason more manufacturers are focusing on CO_2 sensing technologies that warn the diver earlier enough to permit a safe bailout.

CHAPTER 8 - FAILURE MODES AND EMERGENCIES

Carbon dioxide breakthrough is considered a catastrophic loop failure. You cannot safely operate in CCR or SCR mode and must switch to an open circuit gas supply immediately. This may be provided by your BOV or through a bailout tank with second stage.

Given the scenarios above, it is imperative that you carry adequate bailout gas to get yourself safely home. As mentioned previously, in a few cases of extreme exploration my team has opted to share bailout beyond a certain point of penetration. I find that carrying two, 80 cubic foot tanks in a sidemount style is very easy and comfortable and will get me out of most dives. I use smaller tanks for lighter penetration or easier dives. Beyond that, staged gas is preferred. Understanding that carbon dioxide issues will significantly elevate the SAC rate of an exiting diver, I am very conservative with gas planning. In some reports, a diver's SAC rate easily doubles and does not recover to a normal level during bailout.

Some instructors recommend a bailout strategy that requires a dive team to carry a minimum of 1.5 times enough gas to get a single diver safely to the surface. This might be adequate for some people's risk assessment, but in a team of three cave divers, I find this to be too lean. In this scenario, there is no single diver that can make it home on their own gas. This could force a team of three to stay close and swap tanks throughout the exit so that a diver is never left without open circuit gas (which may also supply a drysuit) of their own. Other divers have taken drastically different approaches to planning bailout. Some advocate a one-hour rule that gets any diver topside with one hour of any consumable in reserve, whether batteries, gas, scrubber or bailout.

During your class you will learn how to remove and replace bailout tanks and comfortably swap with another diver. When an OC tank is half depleted, you should switch with a diver who has a full one. Everyone should be carrying a tank with some bailout gas in it, not carrying around empty cylinders.

My best advice is to make a careful risk assessment prior to your dive and visualize the worst-case possible scenario in that environment. Only then, can you make the right decision for you and your team about the amount of gas to carry in reserve.

CHAPTER 8 - FAILURE MODES AND EMERGENCIES

Manual Control

If your solenoid valve fails or if you are operating an mCCR, you will be adding oxygen to the loop via a manual injector button. As stated earlier, many eCCR divers opt to use manual control from time to time to manage bottom time or ascents. You'll practice running your unit with manual control during dives and ascents and even for entire dive profiles to ensure that you are comfortable with the skill and familiar with the pace of injections to maintain PO_2.

Gas Supply Failures

As with any dive gear, there may be a time when a hose unexpectedly ruptures or when some other issue causes rapid gas supply loss. Fortunately these failures are accompanied with a "boom-hiss" most of the time.

mCCR divers will add their own oxygen by watching their displays. eCCR users will learn and practice this skill often so they are completely adept at handling solenoid failures and other incidents that could force them into manual operations.

On some rebreathers, rapid depletion of a gas supply will cause a "high rate" alarm. If this happens to you on a CCR, you will immediately shut down the gas supply and start trying to figure out where the leak is coming from. In the worst case scenario you can feather the tank valve hand wheel on and off as needed to protect what is left. In a perfect world, you actually don't need diluent to get back to the surface. You only need a little bit of oxygen to bring up the PO_2. You should practice orally inflating your wing to be sure you can reach the mouthpiece and achieve positive buoyancy once you are up on the surface.

Regulator Failure

Sometimes first stages fail. They are designed to be failsafe and should result in a free flow. Some rebreathers will warn of this situation and others may not. You may simply see your PO_2 rising above setpoint if your oxygen regulator fails. In all likelihood, you will also ascend slightly from the additional gas volume. In this case, if you are a technical diver, once you have determined the problem, you can turn the tank valve hand wheel off, then feather it on slightly whenever you need to bring up the PO_2.

Mask Clearing

A task as simple as clearing your mask adds new complexity on a rebreather. When you vent gas into your mask, you are depleting the loop volume. That will

CHAPTER 8 - FAILURE MODES AND EMERGENCIES

cause you to sink. A small low profile mask is preferred so that minimal gas will be needed for clearing. Your mask should fit well so you don't have to do this often. When you are ready to clear your mask prepare for a slight loss in buoyancy. If the ADV fires diluent into the system, then the resulting dilution can cause the solenoid to fire and bring you up in the water column. Some adjustment may be needed to loop volume, buoyancy and PO$_2$ after mask clearing.

Open Circuit Bailout

Open circuit bailout to the surface is one of the most important skills to master on any rebreather. Recreational rebreathers rely on OC bailout as the primary or only action in the event of a failure. Whether you are diving a recreational or technical rig, you should be comfortable making the switch from closed circuit to open circuit.

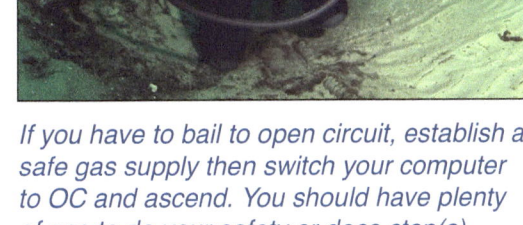

- ✓ Identify the problem.

- ✓ Switch the DSV lever to open circuit to secure the loop.

- ✓ If you have a BOV with an adequate gas supply, keep breathing on the BOV.

- ✓ If you do not have enough gas or if the BOV was contaminated in a flood or damaged, then switch to the alternate second stage.

If you have to bail to open circuit, establish a safe gas supply then switch your computer to OC and ascend. You should have plenty of gas to do your safety or deco stop(s).

- ✓ Confirm with your buddy that you are ready and begin your ascent.

- ✓ Switch your rebreather or computer to OC mode to track decompression if applicable.

- ✓ Do a safety stop or decompression stops as required.

- ✓ Monitor gas supplies closely and prepare to share if needed.

You need to be comfortable sharing gas with another diver from your bailout tank or asking another diver to use their second stage for ascent. All scenarios should be practiced. If you are assisting another diver with a second stage, bear in mind that they will need to orally inflate their wing on the surface if they have run out of gas. Standby to offer assistance until they have achieved positive buoyancy. Many drownings occur on the surface when a diver involuntarily aspirates water.

SCR Mode

Any rebreather can be operated in SCR mode. Depending on your unit and instructor, you will attempt various iterations of this skill in class. If you are short

CHAPTER 8 - FAILURE MODES AND EMERGENCIES

of bailout gas to get you out of an overhead environment and have either lost your oxygen supply or electronics, this skill can significantly extend your gas supply. The deeper you are, the more effective the gas savings of the drill can be. If you know and trust your computer readings, watch your handsets during this drill. When your actual PO_2 drops about a tenth below your calculated diluent PO_2 (at 100 feet/ 30m on air, your diluent PO_2 should be .84) then vent a breath and refill the volume with diluent. Keep doing this as you see the PO_2 drop. In the end you may be venting every fourth or fifth breath or greater then topping up the volume with diluent each time.

It is critical that you become completely familiar with the features of your handset. Confirm that everything is set correctly before beginning your dive.

If you have had the worst day possible and have no electronics, no oxygen and not enough bailout and diluent gas to get home you can use this skill to extend your gas supply as you move horizontally. Any time you make an ascent, you will need to flush to confirm diluent PO_2 and recalculate in your head what that might be. The shallower you get, the riskier this move becomes. Your diluent PO_2 after a flush at a shallow depth can drop to hypoxic levels quickly. If you have saved gas on the way out of the cave or wreck, hopefully you will have enough to breathe on open circuit at shallower depths and on ascent. If you still have displays, then you can easily monitor and continue in SCR mode.

SAFETY CHECK

Carbon dioxide buildup is not resolved in SCR Mode. Using SCR in the event of a CO_2 issue may lead to death.

Various Electronic Failures and Features

Depending on the complexity of your rebreather, there may be a lot of skills involving very specific manipulations of the electronics. These types of features are only seen on technical units and are often intended for long range dives when you have a lengthy exit through a wreck, cave or decompression ceiling.

Some features include:

- **Disabling a cell-** used when a diver has confirmed a single errant sensor and wants to suppress the alarm to prevent alarm fatigue on abort (when an alarm

CHAPTER 8 - FAILURE MODES AND EMERGENCIES

is repeated and is unable to be turned off, then a subsequent alarm may be masked or ignored).
- **Disabling a pressure sensor**- used when a tank pressure sensor has failed.
- **Disabling a thermal array**- Thermal Profile Monitors or Tempstiks have occasionally failed. Some rebreathers permit the user to disable the alarm to prevent alarm fatigue on exit.
- **Switching decompression algorithms**- this features lets a diver at the surface switch between several different algorithms for the purpose of changing the conservatism factor of their dive profile or closely matching a dive partner or backup device. (Only used when completely clear).
- **Diluent PO$_2$**- allows the diver to quickly find the expected diluent PO$_2$ at depth without having to make the calculation in their head when conducting a diluent flush. Some rebreathers use this figure to warn you if you go too deep for your diluent.

CCR Instructor Peter Sotis walks a diver through the features of his handset during a try-dive at Rebreather Forum 3.0 in Orlando.

- **Switch to OC and Return to CC**- this should be well practiced and familiar for aborted dives.
- **Master Error Screen**- Some computers offer a prioritized list of alarms when multiple failures occur at once.
- **Track PO$_2$ or SCR Mode**- Some computers can be set to SCR mode so the unit tracks the diluent PO$_2$ for the purposes of decompression.
- **Battery Warnings**- Most handsets will warn of depleting batteries. Some move into power-saving mode that isolates important safety measures from less necessary functions. For instance, some rebreathers will darken the screen to save power for solenoid operation as batteries are depleted. Some will gently pulse a dim screen to attract attention and save power. You should get familiar with your specific features since you are unlikely to experience them first hand until you have an issue.
- **Simulators**- some handsets are equipped with simulators that permit you to see what the screen looks like underwater as well as assisting with followup dive planning.
- **Multiple OC and CC gases**- you may be able to input several OC and CC gases in your computer so that they are quickly switchable on a complex dive.

Understanding and practice of the failure modes and procedures specific to your rebreather is key to safe operations.

-

9

Rescue

In this chapter:
- *Medical Emergencies*
- *Decompression Emergencies*
- *Recovery Operations*

Rescue

Medical Emergencies

Rebreather diving can be very strenuous. Divers with underlying medical issues may suffer from a variety of conditions that can result in very real emergencies for the entire team. Heavy workloads can contribute to heart failure or strokes. High partial pressures of oxygen can lead to seizure, convulsions and drowning. Prescription medications may contribute to seizure activity or unconsciousness in otherwise reasonably healthy individuals. Improper gas mixtures can cause unconsciousness from narcosis, hyperoxia or hypoxia. Rebreather dives carry risks of carbon dioxide poisoning, hypoxia, hyperoxia and caustic burn injuries caused by wet scrubber materials. There are numerous situations that can occur well into a dive, leaving the team in a position to facilitate a rescue.

Assisting an Unconscious Diver on the Surface

Unconscious divers on the surface should be given immediate, positive buoyancy and their airway should be protected from water. If the diver is close to the water's edge, prioritize getting them onto the boat or shore and into a position where you can "look, listen and feel" if they are breathing. If they are not breathing, then immediately give two full rescue breaths while pinching the nostrils and keeping the airway open. Get into a position where you can check for pulse. If they do not have a pulse, then begin rapid and continuous CPR compressions. Activate the emergency medical system (EMS) as quickly as possible and give oxygen if available. Continue to provide basic life support (BLS) through CPR and rescue breathing until an ambulance or other medical personnel arrive to assist.

CHAPTER 9 - RESCUE

Unconscious Diver Underwater

Very few divers who are unconscious and non-breathing in an overhead environment or at extreme depths will survive. A few such open water divers have been successfully rescued. With this in mind, a rescuer must always consider their own safety during any rescue attempt. If the loop is still in the diver's mouth, try to keep it there to protect the airway. Flush with appropriate diluent if you suspect an issue with PO_2. Close the loop to prevent excessive negative buoyancy due to flooding. If the loop is equipped with a BOV, then the closed loop ensures a safe breathing mix as long as it is not hypoxic. If the loop has fallen out, do not replace it since it will be flooded. If multiple rescuers are involved it may be possible to get an open circuit regulator into the diver's mouth. On the surface, establish buoyancy, get them to shore or the boat, begin basic life support and activate EMS.

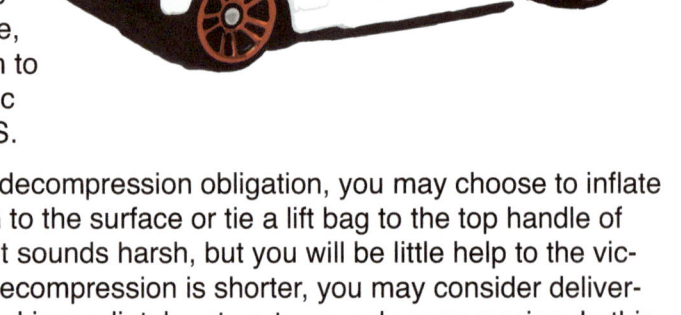

If you are under a lengthy decompression obligation, you may choose to inflate the victim's wing to send him to the surface or tie a lift bag to the top handle of his rebreather and inflate it. It sounds harsh, but you will be little help to the victim if you are injured too. If decompression is shorter, you may consider delivering him to surface support and immediately return to your decompression. In this case you should repeat some of the deeper stops and/or lengthen your entire decompression, as able.

These choices are gut wrenching, since they inevitably involve a friend or even a life partner. These sorts of emergencies should be discussed before diving and must be a part of a full risk assessment and dive plan.

Since first aid recommendations are revised from time to time, you should remain current in your first aid and CPR training.

Decompression Emergencies

Omitted Decompression

If a diver misses some of their decompression and surfaces early, there are a number of procedures that different organizations have developed as standard protocols. The U.S. Navy, Comex and other commercial diving operations and military units have determined that, if the diver is asymptomatic, and can return to the water within five minutes, there are many options available. In some cases, the stops are repeated and lengthened by 1.5 times. In other cases, the

CHAPTER 9 - RESCUE

diver descends one stop deeper than the stop she aborted and adds time all the way up. Several methods have proven successful, but are beyond the scope of this text. Whether the diver repeats stops, lengthens stops or stays out of the water, breathing oxygen after the dive will be beneficial. Even if the diver is asymptomatic, prophylactic oxygen may prevent a full blown case of DCI.

Decompression Illness

Decompression Illness (DCI) is a very real risk for technical divers. Many minor hits are not reported or treated, and there is still an unfortunate atmosphere of secrecy surrounding DCI. Divers are embarrassed, and feel that getting bent reflects on their ability as a diver. This could not be further from the truth. Technical divers push the edge of the envelope every day. Decompression tables are only theoretical mathematical models; fitness, age, hydration and other factors may increase the likelihood of getting bent. Preventive measures may lessen your odds of getting bent, but there are no guarantees in diving.

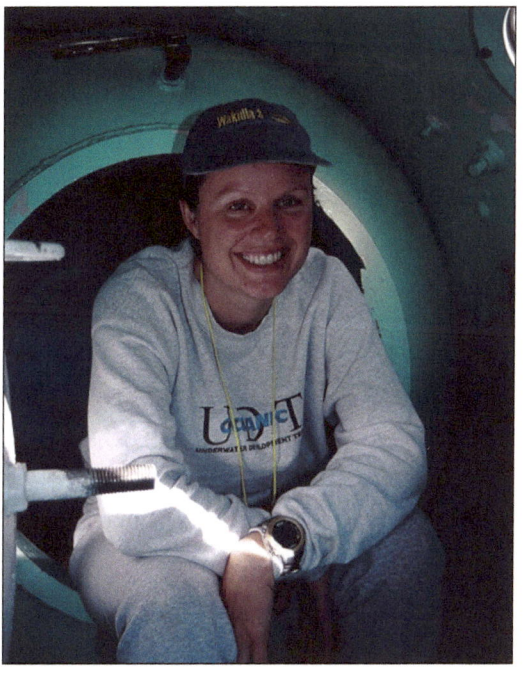

Emerging from a recompression chamber used as a decompression habitat.

Decompression Illness is a term that describes both decompression sickness (DCS) and arterial gas embolism (AGE). DCI results from inert gas bubbles that grow within tissues, causing localized damage of one form or another. AGE happens when bubbles are injected into lung circulation, then travel through blood vessels, causing a blockage in blood flow. Both ailments are treated the same at the level of first aid and are thus often discussed in tandem. Both injuries result from a reduction in pressure surrounding a diver.

The longer you stay down and the deeper you dive, the more inert gas, such as nitrogen, is stored in your tissues. The inert gas does not serve the body in any positive way, but oxygen molecules are used up as fuel. As long as you surface slowly, and according to a previously determined mathematical model, then in most cases, the extra inert gas will be slowly off-gassed as you come up. If you surface too quickly, or have other contributing factors, then the inert gas will come out of solution, forming tiny bubbles in your tissues. Bubbles that form in joints, causing pain, are associated with what is known as a "classic bend," but DCI has many other manifestations and complications.

CHAPTER 9 - RESCUE

Treating DCI

Signs and symptoms of DCI usually occur within 15 minutes to 12 hours after surfacing, with a high proportion of serious manifestations occurring within the first hour. Some cases are detected underwater or immediately upon surfacing and rare cases have been reported after lengthy delays, especially after flying.

DCI doesn't always hit a diver like a brick wall. It may creep up slowly and get worse. As such, denial is very common. Minor symptoms may be ignored as insignificant, but it also seems clear that an afflicted diver may not be able to make a reasonable decision for themselves. Denial may be a symptom of chemical changes that are happening within the body and brain. It will be incumbent upon the dive buddy to take charge of the situation and arrange for proper treatment.

The only definitive treatment for DCI is recompression at a hyperbaric facility, though first aid can greatly improve the outcome for a diver. In some minor cases, oxygen may resolve symptoms completely, but a medical professional should make that judgment, since oxygen first aid may only delay symptoms for a while before they recur in a similar fashion or get worse.

Immediately place a diver with suspected DCI on oxygen and make a rapid assessment of the urgency of his condition. Oxygen should be provided at the highest concentration possible. Using a demand valve is preferred, but a continuous flow mask with flow rate of 10 to 15 liters per minute may be used if supplies are abundant. If you do not have an emergency oxygen kit, use your rebreather as a substitute if the diver can sit up comfortably and breathe from the loop. In this case, pinch his nose or have him wear a mask to optimize supplies and PO_2.

When symptoms are severe and onset is rapid, EMS must be activated immediately, since these patients will need to be medically stabilized before recompression therapy can begin. Divers Alert Network (DAN) should be notified early to link emergency physicians with hyperbaric specialists. Severe signs and symptoms may include:

CHAPTER 9 - RESCUE

- Severe and rapid onset of symptoms within an hour of surfacing
- Unconsciousness
- Worsening condition
- Difficulty breathing, dizziness
- Coughing
- Paralysis
- Stumbling and weakness

Careful monitoring of airway, breathing and circulation is important with an understanding that vomiting may occur. Lay the diver on her side unless CPR is being administered or if using a rebreather to provide oxygen. If aircraft evacuation is required, cabin pressure should be limited to as close to sea level as possible, so it does not exacerbate symptoms.

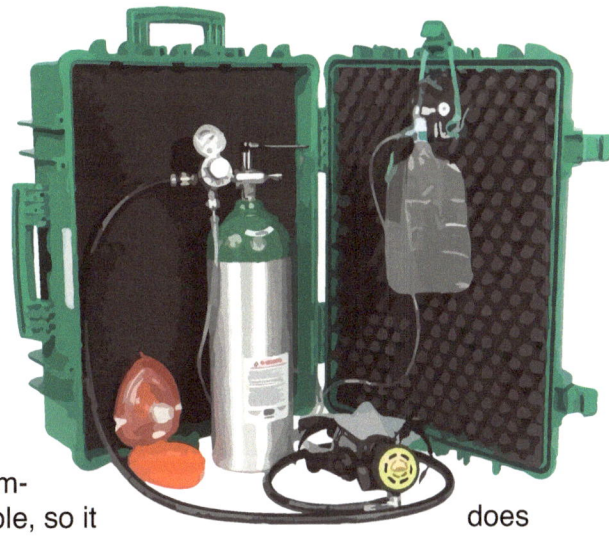

When the diver arrives at a hospital, medical staff will attempt to assess and stabilize the victim. IVs are often established to rehydrate the diver and urinary catheters may be inserted. After stabilization in an Emergency Room, DAN staff help organize evacuation to the nearest chamber facility. They contact the receiving chamber and relay medical information so that staff may be advised to prepare the hyperbaric facility. Many chambers operate part time. Others are active in medical treatments around the clock. DAN will find the closest option for rapid, appropriate therapy. They also stand by for diagnostic assistance, and if necessary, make their own facilities available at Duke University Medical Center.

If you are accompanying your dive partner through medical evacuation, gather all the information you can about her dive, medical history and onset of symptoms. If you know how to conduct a field neurological test, do so, and continue to watch for changes in her status.

Many cases of DCI are less severe and may not require emergency transport. In these cases, the diver may be experiencing severe pain, especially in joints. The pain may be getting worse, but is progressing slowly over a period of hours. These divers should be immediately placed on oxygen and generous fluids by mouth to aid in rehydration. Calling DAN will help to make the decision about whether you can transport the victim or whether more urgent evacuation is needed. DAN will stay in contact as you travel to ensure that worsening symptoms do not require rapid intervention. The diver will, in most cases, still need to go to an emergency room for clearance, before being sent to a hyperbaric chamber. In some parts of the world, divers are accepted directly at remote

CHAPTER 9 - RESCUE

chambers, but in the United States, medical clearance in an ER will usually come first.

As your dive buddy's advocate, you should connect DAN with the emergency room personnel and let administrative clerks know that even though your dive buddy may be walking and talking, their condition requires urgent attention. Once in the emergency room, a patient will be given an EKG and chest x-ray to ensure he is not having a heart attack and will likely be placed on intravenous fluids with continued oxygen until he reaches a chamber.

Some cases of DCI may be subtle. Signs and symptoms may appear vague and present themselves over a period of many hours or even days. Many divers ignore symptoms for several hours, reporting their situation only when they are unable to sleep at night. That is why many chamber treatments are organized by on-call staff in the early hours of the morning.

Technical diving can require a lot of physical effort both before and after a dive. Being too active after a dive could contribute to factors leading to decompression illness. Take time to relax and pace yourself and avoid overexertion.

In mild cases, you may not even be sure if you have DCI. Call DAN for advice. Report all diving activities for the previous two days. Describe all signs and symptoms, which may include blotchy skin, rashes, pain, numbness, unusual niggles or other sensations. DAN will make a judgment call and likely advise you to go to a local medical center for evaluation. Hydration is important in all cases.

Leaving DCI untreated is unfortunately widespread in the technical diving community. Whether it is ignorance, embarrassment or denial, active technical divers will tell you that they know many divers who have ignored symptoms and resisted treatment all together. If the DCI is severe and untreated, a diver may be left with long term disabilities, dysfunctions or muscle weakness. Spinal cord injuries can also result. More subtle cases involving joint pain, may eventually contribute to arthritic conditions and degenerative disease like osteonecrosis.

Factors that May Contribute to DCI

There are numerous factors that may predispose an individual to DCI. Most of these factors remind us of the importance of general health and fitness, especially cardiovascular. Dehydration is noted in most cases to some degree but is also a *result* of the illness. Dehydration causes fluid loss in tissues as well as electrolyte imbalance. It may be caused by excessive sweating in warm climates, breathing dry gases, immersion diuresis, intentional dehydration to permit drysuit diving and general lack of fluid intake many days before a dive. Good hydration benefits general health and may help prevent DCI. Other contributory factors include age, injury, illness, obesity, previous DCI hits, Patent Foramen Ovale (PFO), rapid ascent, omitted decompression, omitted safety stops, travel to altitude, consumption of alcohol or drugs and heavy exertion during or after a dive.

Bent

It is said that good judgment is born from experience. Yet, unfortunately, experience often comes from bad judgment. Over a decade ago, I earned a lot of experience when Paul Heinerth and I eagerly hiked into the steamy Yucatan jungle to explore deep passages of The Pit; a series of deep passages we uncovered three years earlier. Armed with newly developed rebreathers and precious helium mixes, we set out to explore the mysteries of the tunnels in this popular region of Mexico. We'd be diving nearly 400 feet deep.

Today, divers flock to The Pit by an easily accessible jungle road, but in the first exploration efforts of the Sistema Dos Ojos, getting your gear to the dive site required exhausting hikes, makeshift donkey carts, robust Mayan sherpas and a lot of determination and sweat. Paul and I had been anticipating the help of a local support team, but on our arrival discovered they had been otherwise dispatched to assist the newly thriving recreational tourism business. The shop was simply too busy to free up any of their divers to help us. Knowing a hundred divers that would have eagerly joined us from the U.S., I was a little perturbed, but resolved that we could do it on our own. We had plenty of equipment, all our expedition supplies and a trusty Comex In-water Recompression Table.

Our logbooks were fat with many thousands of dives and perhaps I was feeling a little young and invincible. I was keen to quickly bag as much new exploration as possible. Some combination of passion and spirit led me to make a decision to dive very deep, two days in a row, something I would not have done earlier in my career and something I have not done since that punishing day.

Feeling marching ants crawling under the skin of my thighs at 60 feet of depth during decompression, it was clear to me that I was bent. I had never felt a sensation like that before. I lengthened my decompression profile and gingerly climbed out of the water, but the profound exhaustion told me I still had a problem. Too tired to do anything other than breathe

CHAPTER 9 - RESCUE

oxygen and drink water, I collapsed on my sleeping mat and hoped everything would go away. My resolve was eroded. I lost the ability to help myself. I was scared.

Improvised in-water recompression at the dive site, an arduous hike out of the jungle, and three followup chamber runs resulted in a full, but painful recovery. Looking back on my precarious situation, I learned many things. I believe that the frequently reported symptom of "denial" extends far beyond embarrassment. I'm convinced that a bent, sick person knows what to do, but simply can't do it. To me, it makes sense that if your body chemistry is screwed up, so is your ability to make a decision to help yourself. It is therefore critical that somebody on your team takes charge and makes the best decisions for a DCI victim. Secondly, oxygen is a very powerful ally. Having plenty of rich gas on site saved me from grave injuries. Lastly, the long-term psychological effects after a DCI incident are worthy of discussion. Our community has progressively discarded terminology that divides decompression illness into "deserved" and "undeserved" hits. DCI, like a football player's concussion, or a runner's pulled hamstring, is a sports' injury. When we share our experiences, divers will be better informed if it happens to them. Do I possess better judgment and have I gained wisdom and experience? Yes.

Am I ever going to get bent again? I can't say, but I do everything I can to minimize the risk.

Patent Foramen Ovale (PFO)

A PFO is an arterial-septal defect, which allows blood to shunt from one side of the heart to the other, bypassing lung filtration. When blood shunts from the right to the left side, it may carry micro-bubbles with it. These micro-bubbles can enlarge and pass directly into arterial circulation.

During gestation, a natural hole links the arterial chambers of the heart in the fetus. After birth, that hole closes and heals when a baby begins to use its lungs. In up to one-third of adults, the flap does not heal fully and blood will shunt when increased pressure from coughing or exertion occurs. In diving, this allows venous blood, perhaps filled with micro-bubbles, to return directly to arterial circulation, bypassing the lung filtration and the chance to off-gas through normal breathing. Once in the arteries, bubbles may gang up and grow in size. These bubbles are more likely to cause neurological symptoms. They may block circulation to parts of the brain causing stroke, may lodge in the spinal cord causing paralysis, or may lead to heart attack.

Specialized physicians may test for PFOs by injecting gas directly into the circulation. Asking the diver to conduct a forceful Valsalva maneuver (ear clearing with pinched nose), they can look for bubbles on an echo-cardiogram. They may also ask the patient to do deep knee bends to further raise pressure that might reveal bubbles crossing the septum.

Experts do not suggest that divers seek prophylactic testing for PFO, and many divers with aggressive diving activity and a PFO never get bent. That being said, given that up to one-third of divers may have some degree of shunting, it seems prudent to be very conservative about post-dive exertion that can raise pressure and lead to an opportunity for bubbles to move into the arterial circulation. Specialists, such as technical diver and anesthesiologist, Dr. Simon Mitchell, suggest that the event rate of severe DCI is still very low in people who have a PFO. However, if you experience neurological DCI or frequent cases of skin bends, you may consider consulting a specialist that can discuss risks and benefits of testing.

After a dive involving decompression, take time to do some extra surface decompression, relaxing after your dive on the surface. Try to avoid heavy exertion such as climbing a ladder while wearing all your tanks and CCR. Pass bailout tanks up to a Divemaster or hang them on a gear line for later retrieval.

Physician's Clearance

If you are taking any regular medications, discuss these with your doctor. Ensure your doctor knows you are a diver. You can also consult DAN about the risks of certain medications. They may seem like common medications to you, but your Viagra or Sudafed could change your blood pressure, heart rate or seizure susceptibility. It is always best to ensure that an expert who is familiar with your conditions and medications clears you for rebreather diving.

An Important Investment

DAN Insurance is extremely inexpensive. Having personally used their services, I cannot say enough about the excellence of this organization. You simply must purchase DAN insurance or its equivalent. There are several different levels

CHAPTER 9 - RESCUE

available from basic coverage to extensive additional travel insurance. Their website and magazine *Alert Diver* will keep you current regarding medical research. DAN's hotline is always available to answer non-emergency questions and organize treatment if needed.

The DAN Emergency Hotline number is +1-919-684-8111 or +1-919-684-4DAN. You will be connected with an expert in diving medicine. They accept collect calls from anywhere in the world and answer calls 24 hours a day.

Recovery Operations

I questioned whether I should go into detail regarding recovery operations in an entry level rebreather book. I decided it was important for you to know what happens in the case of an accident. I have grieved over many departed friends over the years and wondered about whether we should have been talking about more of the gory details so people don't bypass risk assessment. I also think it is important for the future of accident analysis that we better understand why this form of diving seems to be about ten times more likely to end in an accident than other types.

There are very few successful rescue operations in technical diving. In most cases, recovery operations are initiated to bring the body and equipment back from depth. In a recovery, the most important thing to remember is that the situation is no longer urgent. Care must be taken to protect others from harm. Patience will allow specialists to gather important information that may assist in accident analysis. Local law enforcement should be notified of the importance of protecting the evidence that may be retrievable if a rebreather is handled properly and the potential for loss of evidence if it is not.

If you are on site when a diver is overdue, there are many important functions that you can fulfill. Search teams can be dispatched. Written records, with careful time annotations should be started. Dive partners may be interviewed for important information including time, depth and last location at which the diver was seen. Emergency equipment can be gathered at the water's edge in case it is needed. Backup tanks can be organized in case a diver returns with insufficient supply to complete an open circuit decompression.

On a long term operation, an entire infrastructure may need to be built for the comfort and safety of search and recovery divers. Food, fluids, warmth and protection from the weather will all be important. Accident reports and witness statements should be provided to law enforcement as soon as possible to expedite the progress of accident investigation.

The state of the victim's dive gear is extremely important in helping to determine the root cause of an accident. Trained recovery divers will carefully examine gear in place and post-recovery for important clues. Rebreather equipment should only be handled by experienced individuals. Examination of a CCR must

CHAPTER 9 - RESCUE

be timely to be of significance since data can be lost when batteries are depleted. Changing the current state of anything on a victim's equipment is tampering with important evidence that may be used in either exonerating or highlighting equipment liability or operator error. The chain of custody must be carefully documented and include observation by law enforcement personnel.

Underwater Assessment

If a rescue is possible, then all efforts must be taken to save the victim, but if you are a part of a search and recovery team and find the victim deceased, there are some important things that should be noted.

Is there anything in the diver's mouth?

➡ rebreather loop (open or closed loop)

➡ BOV

➡ second stage from bailout tank

➡ nothing

➡ is a second stage dangling?

In what position is the switch on the DSV?

➡ horizontal

➡ vertical

➡ all the way up

➡ all the way down

➡ somewhere in between positions

✳ if you close the loop for ascent, note it in your report

Sketch out everything that you see on a handset in position. If there is a secondary display, do the same.

➡ was the handset worn on a wrist?

➡ was the handset dangling?

➡ if there are slider switches on a handset, what position are they in?

➡ is there a flashing HUD and if so, what color? Are any lights or screens flashing?

➡ Do you hear any beeping?

➡ Do you see any bubbles?

✳ If you encounter a blank screen push one button and record what you see. Push the second button and record what you see.

Note the victim's body position.

CHAPTER 9 - RESCUE

➡ What color is their face?

What state of buoyancy are they in?

➡ Are they on the bottom?

➡ Are they neutral in the water column?

➡ Were they found positively buoyant on the surface?

Note any missing equipment.

➡ Is the mask present and if so, was there fluid in the mask?

Note the tank pressure on each SPG (before surfacing if possible).

✷ Do not touch tank valves.

When taking the victim up, did the BC inflator work?

Make as many observations as possible underwater while there are no external pressures to rush. As soon as you reach the surface, things get rather chaotic. The body and equipment may be harmed in the effort of recovery and evidence can be lost.

Surface Assessment

Once the body has been recovered, it is important that law enforcement takes the lead in investigating the accident. If you can, try to recover the victim in their gear in one unit. Make sure law enforcement personnel know what they can grab and what they shouldn't pull on. Don't let them try to heave the victim or the gear on the boat by the breathing loop or other delicate devices.

If they are amenable, you can assist with a brief assessment of any equipment that you are familiar with. At this stage, the assessment must be completely non-destructive. Anything that is examined should be left in the position or condition in which it was found. If you are not familiar with the rebreather, there are only a few things you can do without potentially harming evidence. Urge law enforcement to bring in an expert right away.

Photography, though gruesome, is a helpful tool in accident analysis. It is essential that every single step of an assessment is photographed and attended by law enforcement. Don't just jump in and start grabbing things without permission or you may be tampering with evidence.

The first step should ideally be taken with the victim in their equipment if possible. If not, note as much as possible prior to gear removal.

It is essential to determine every connected hose, every switch position, which side bailout tanks were worn on, which clips held the tank, whether the diver had a DSMB and reel and whether it was deployed, which hose fed the dry suit, etc. You may not know the significance of details right away and every detail may be

CHAPTER 9 - RESCUE

important in determining root cause(s) of an accident. Every item should be photographed in multiple angles.

Detailed note taking is essential. Everything should be recorded including items that appear to be functioning properly or connected properly. It you noticed that the bailout tank was properly feeding the ADV, write it down. If you get to something such as fins and they look fine, write down that there were no faults found with the diver's fins. Inventory every single piece of equipment that is stripped from the victim. Photograph each piece and note everything. Keep all equipment together and surrender custody to law enforcement. Even if they don't think the fins are significant, everything should be secured together.

Prevent all unauthorized personnel from having contact with the diver's equipment. Immediately contact the manufacturer to confirm the safe powerdown procedure for battery and computer verification to retain the best evidence.

Witnesses

Witness statements are critical. Each witness, whether with the victim or just on the boat, should give a statement about what they saw. Even people who view an assessment can offer valuable insights into the accident or verify a chain of custody of gear. Witness statements can also immediately rule out certain problems. For instance, if the diver passed out on the boat prior to the dive, then hyperoxia can be ruled out.

Functional Assessment

A complete functional assessment should be made as soon as possible after the accident. Un-rinsed gear sitting in a morgue deteriorates very quickly and evidence is lost. Batteries die. Salt seizes valves, switches stick and flooded canisters can quickly corrode electronics and sensors. Ensure that the police know how important it is to immediately identify an expert for examination. Make certain that they understand that the material in a flooded loop may cause burns. Though it might seem counter intuitive, immediately contact the rebreather manufacturer. They have a great interest in verifying that the unit is properly assessed and will want to be present for any examination. Though they will not physically touch the gear during an investigation, they can alert examiners about potential damage to evidence and point out information that can be gathered. They will also have the means to download and analyze onboard data. When a Boeing Aircraft crashes, you can be sure that the manufacturer is involved in accident analysis. A rebreather accident should be handled in a similar fashion.

Many rebreather manufacturers post functional assessment and safe shutdown procedures on their website. They can be easily accessed before making any improper moves that could destroy evidence.

CHAPTER 9 - RESCUE

Hyperbaric Assessment

There are several test labs in the world that are equipped with qualified and calibrated gear used to test rebreathers under hyperbaric conditions that are similar to the human body. They are capable of challenging carbon dioxide scrubbers to determine whether CO_2 may have played a part in an accident. A partial list of qualified labs can be found on the AP Valves website and is included here:

USA

Navy Experimental Diving Unit:
http://www.supsalv.org/nedu/nedu.htm
Centre for Research in Special Environments (CRESE), Buffalo University:
Contact Prof. Lundgren- Tel: 1 (716) 829-2310

Europe

ANSTI Test Systems Ltd: Tel: +44 1489 575228
Contact Ian Himmens:
Ian@ansti.com
Health & Safety Laboratory:
http://www.hsl.gov.uk, Tel 01298 218334
nicholas.bailey@hsl.gov.uk
QinetiQ: Contact: Gavin Anthony, Tel: +44-(0)-23-9233-5146
Email: tganthony@QinetiQ.com
Ambient Pressure Diving Ltd:
www.apdiving.com, Tel: +44 (0)1326 563834
Contact: Martin Parker- martinparker@apdiving.com
Swedish Department of Defence Medicine:
Swedish Defence Research Agency, Stockholm, Sweden
http://www.foi.se

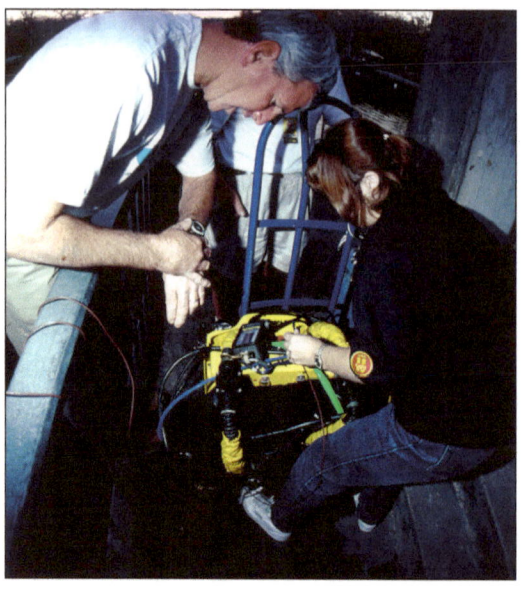

Many eCCRs contain a "black box" that tracks parameters throughout the dive. Data can be retrieved through a download cable or wireless transmission. In many cases an accident can be examined second by second.

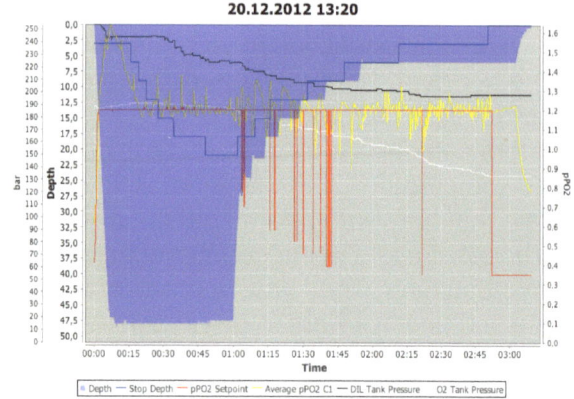

CHAPTER 9 - RESCUE

Careful examination of a rebreather helps investigators get to the root cause of an accident. The coroner will almost always determine drowning as the cause of death, but by looking at equipment, we can sometimes uncover scenarios that developed throughout a dive profile. Those developments, behaviors and faults help us to understand *why* a diver became unconscious and drowned. Accident analysis is key to helping us develop better equipment, improve the human-machine interface and nurturing a safer culture of rebreather diving.

> **SAFETY CHECK**
>
> *Help DAN learn from incidents:*
> *http://www.diversalertnetwork.org/IncidentReport/*

The Wet Mules Dive Team members Dr. Richard Harris, John Dalla-Zuanna and Craig Challen prepare to drop on a deep wall in the Indian Ocean.

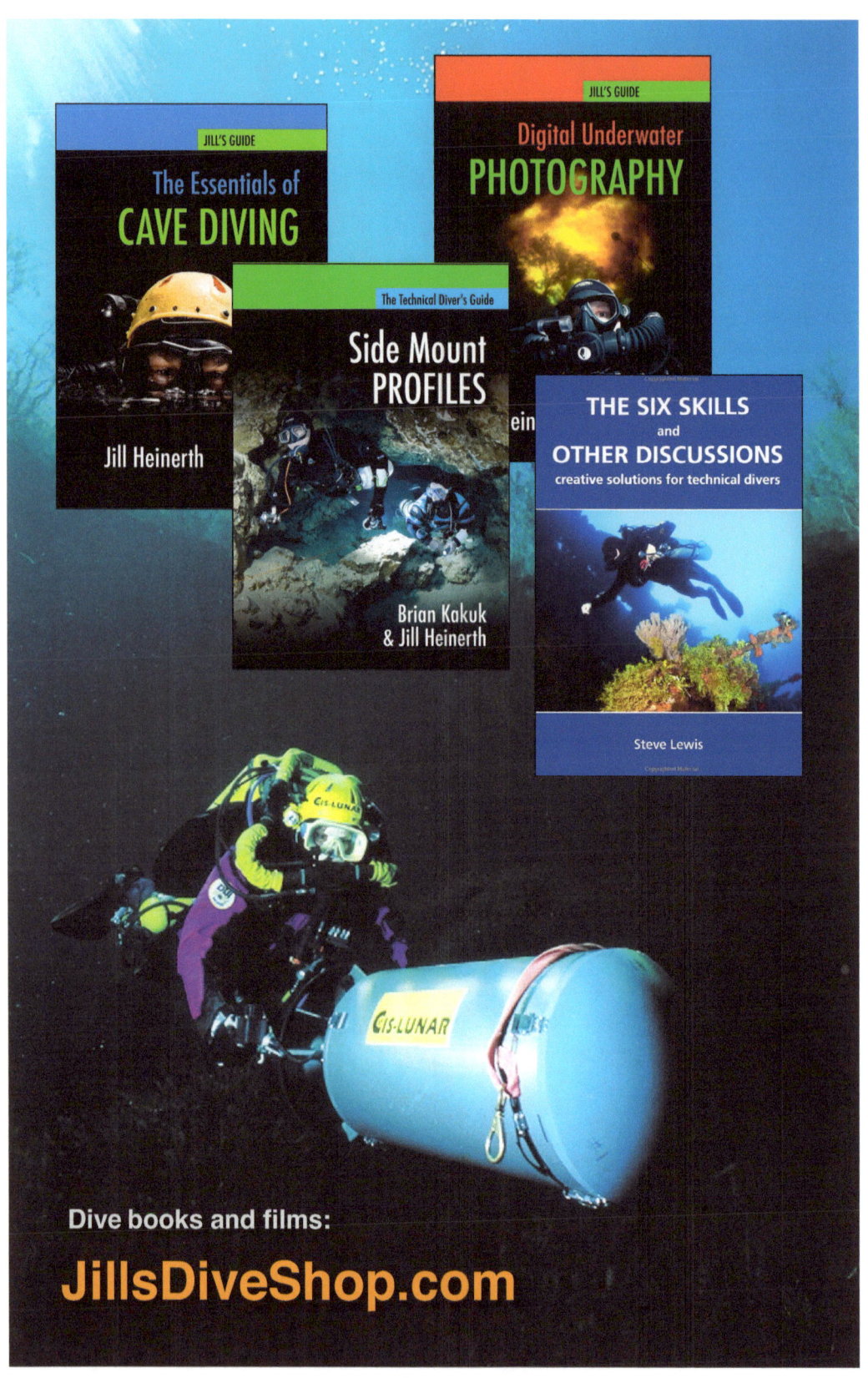

10

Testing

In this chapter:
- *Testing and Validation*
- *CE Testing*

Testing and Validation

Rebreather diving emerged from the garages of tinkerers interested in emulating or modifying early military and commercial rebreathers which were purpose-built for completely different scenarios than a sport diver is exposed to. At first it was a small developing market, but now it has reached a maturity that demands proper testing and validation. Could you imagine jumping in a new car that was equipped with brakes that were never subjected to testing to a documented safety standard? In America, we demand more testing and validation of our kitchen kettles than our underwater life support equipment.

In my opinion, you should never risk your life using gear that has not been subjected to objective unmanned performance testing. Although some aspects of this testing may be proprietary to a manufacturer, there are certain things that should be shared with prospective customers using a standardized metric that creates a level playing field with the rest of the competition. Fuel consumption tests for cars are calibrated, tested and reported to customers. Your rebreather should be tested for its fuel consumption rate- the scrubber duration based on a calibrated test.

CHAPTER 10 - TESTING AND VALIDATION

In North America, manufacturers have an affirmative duty to warn consumers who use their products. The American National Standards Institute (ANSI) has set forth regulations regarding warnings, symbology, and threat level. Manufacturers that conform to these policies use a series of ANSI compliant marking stickers on their products. Beyond warnings in product manuals, videos and published materials, the actual equipment should be properly labeled. Some manufacturers will require an additional "Assumption of Risk" form to be completed prior to using their products. The instructor becomes an important link in the duty to warn as they ask you to complete this form. After training, your instructor may require you to sign a "Skills Sheet" to confirm your understanding and mastery of skills and have you take a test with product-specific questions that confirms your understanding of specific risks.

CE Testing

The diving industry accepted CE testing of SCUBA regulators many years ago. You might have noticed a CE mark with the number EN250 on one of your first stages. That mark ensures that the regulator has been designed and manufactured to a performance standard that is monitored regularly by an authorized testing facility. For rebreathers a specific standard was developed called CE EN14143:2003 which has been recently replaced by EN14143:2013. If you buy a rebreather in Europe, it must qualify under this life support standard for health and safety. This is not yet the case in the U.S. There are popular rebreathers sold in the U.S. that have not been tested to this standard.

Though the standard may not be perfect, it ensures a certain level of engineering excellence, production quality, life support testing, quality control and management of updates and changes in design and manufacturing. EN14143 includes over 100 separate, documented tests including breathing performance in all temperatures, materials quality in all temperatures, canister duration tests, oxygen tracking and others. These tests are conducted with an objective standard using unmanned techniques. After unmanned testing, the gear is further validated with a number of documented diver trials under a range of conditions. It is expensive to complete all the testing and revisions, and some manufacturers have complained that the requirements are onerous and costly. But I would ask you to consider again whether you would drive a car with untested brakes or even plug in a kettle without UL electrical testing!

A second type of certification required in Europe is called Article 11 testing. This is covered under ISO9001:2008 in the U.S. This essentially means that every year, the same equipment is manufactured, tested and warranted in a consistent manner that meets the described manufacturing specification. It also requires very detailed record keeping such as maintenance of an accurate and updated User Manual, proper notification of consumers regarding upgrades and recalls and detailed notation of quality control, faults and problems.

CHAPTER 10 - TESTING AND VALIDATION

Work of Breathing

You should find out whether your rebreather has been tested and validated. There are several tests that you should inquire about and learn how to read. The first one involves Resistive Work of Breathing (WOB). As noted earlier, this is the test that quantifies how easy or tough it is to push gas through the breathing loop as it passes through mushroom valves and other obstructions. This is important because a high WOB could lead to carbon dioxide issues including retention and excessive workload. This test can reveal poorly designed valves. The test is conducted on a breathing machine that simulates the humidity of a diver's exhaled breath and is conducted at the unit's operating depth. In other words, a sport rebreather will be tested with air diluent at an operating depth of 130 feet/40 meters (in a special chamber) and a technical unit will have additional testing using trimix at 330 feet/100 meters. The unit is tested vertically and horizontally. Hydrostatic work of breathing is also tested to quantify the unit's breathing characteristics in a variety of positions. The unit is rotated through several orientations.

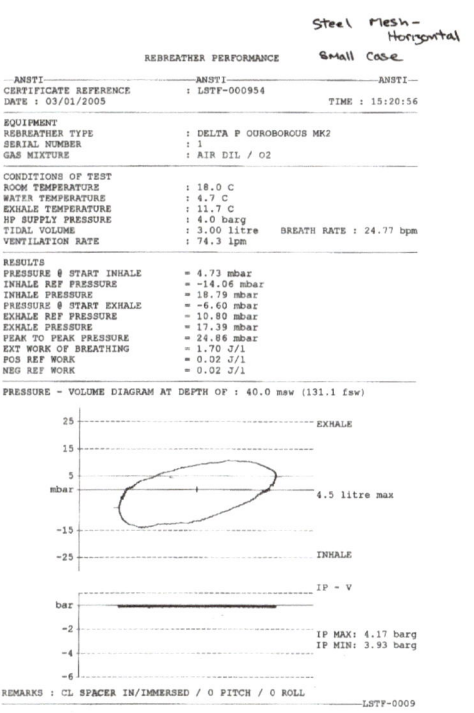

Test results for Work of Breathing.

The WOB results are drafted on a chart with maximum allowed peaks for an inhalation and exhalation of +/- 25 mbar and the maximum effort or workload of 2.75 joules/liter in 4°C/39°F water. The resulting graph looks like a well formed oval slightly tipped on its axis. Ragged jumps in the line could indicate areas of additional resistance such as poorly designed non-return valves.

Canister Duration

Canister duration is also tested at operating depths for the unit and should be provided on a standardized profile. Scrubbers lose performance at depth with denser gas, therefore all rebreathers should be tested using the same parameters of depth, time and gas contents. Check carefully that this is the case if you are comparing one manufacturer's results to another. Some manufacturers have intentionally provided results to consumers using multi-level profiles, warm water

CHAPTER 10 - TESTING AND VALIDATION

or helium rich gas contents that skew results making their device appear better than other units.

In a canister duration test, carbon dioxide is fed into the loop at 1.6 liters/minute (for CE standard). This is a very high carbon dioxide level (and relates to a breathing rate of 40 lpm which can be easily achieved by a working diver), but it assures a safe level and provides a safety margin in case the workload, depth or limits for temperature are exceeded on a dive. The testing is done in 4°C/39°F to offer worst case scenario conditions. The results are reported on a curve and the time is noted when the CO_2 reaches 5 mbar. The time it takes to reach 10 mbar is also noted as an extreme limit. CO_2 alarms are generally keyed to these limits. 5 mbar triggers an alarm alerting the diver to a problem; 10 mbar triggers a second alarm alerting the diver to bailout now.

Sensor Tracking

This test verifies how successful the oxygen sensors are at delivering accurate data over the course of a long profile. It may reveal problems with moisture buildup or other engineering issues. Most CCRs track actual PO_2, but a few rebreathers and most backup computers simply track a given setpoint. A diver assumes that their actual inspired oxygen will be close to the setpoint. The test also checks how well the sensors react to rapid ascents and descents. You want to be sure your PO_2 cannot drop below life sustaining levels on ascent. You also want to know that your sensors don't lag on a fast descent and take you to a peak above 1.6. A descent rate of 100 feet/30meters per minute and an ascent rate of 60 feet or 20 meters per minute are used. Ideally a fast recovery should be achievable bringing the unit back to setpoint.

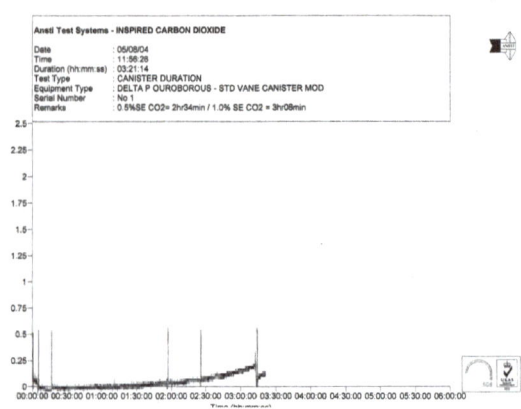

Test results for Canister Duration showing the points when 5 mbar and 10 mbar of carbon dioxide are reached.

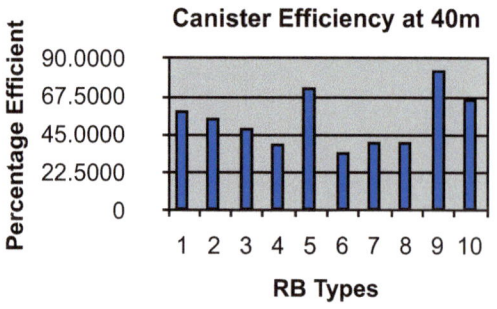

Quantitative canister duration testing is important since it reveals that different rebreather designs net a different efficiency per pound of absorbent material if they are tested using standardized parameters.

CHAPTER 10 - TESTING AND VALIDATION

Other Tests

Other tests include materials testing, pressure testing non-return valves with positive and negative pressure to confirm that the valves can't be sucked through the carrier web, temperature testing of various materials and parts, volume measurements to ensure that dead spaces are not present in the DSV or BOV and others.

Standardized testing and validation provide the consumer with important safety information. It also makes the manufacturing process more transparent as well as compliant with emerging safety standards. Finally it gives a buyer a chance to compare different technologies on a level playing field, revealing good designs and highlighting problems. If a manufacturer is unwilling to share the results of the basic tests above, you should question what they might be hiding. If a manufacturer is unwilling to conduct standardized testing, you should question whether they are dedicated to investing in design safety or are asking you to be a human guinea pig. If they are unwilling to spend the money on testing, they do not value your safety.

Klaus Christiansen and Rasmus Lauritsen emerge from the passages of the Catacombs in Ginnie Springs.

11

Rebreather Myths

We live in the digital information age. We have an endless supply of opinions that can be found at a keystroke. It is a huge benefit to be able to talk to other rebreather divers around the world, but one must also be careful to test the sources of their information. Over the last decade many myths and misconceptions have emerged that continue to circulate in the CCR community.

Myth: Rebreather diving will be safer.

Rebreather diving can provide more options for a diver when they are well equipped with adequate bailout. However, statistics are pointing towards the fact that technical rebreather diving may result in ten times more incidents and accidents than open circuit technical diving. Despite this, most accidents do not seem to be linked to product design. Most accidents appear to be preventable with good diving protocols such as using a checklist, completing a pre-breathe before every dive and never entering the water if your unit fails even a single item on the checklist.

Many experienced rebreather instructors will offer that they become more cautious with years of experience. They understand that our sport is still in its infancy and there is much to learn from researchers and other experienced divers. They also understand the importance of avoiding complacency.

Myth: As an experienced rebreather diver I will be safer.

Again, statistics show that many fatal accidents are happening to experienced rebreather divers. They could be contributed by complacency or might be an issue of risk homeostasis. In other words, "I have survived this, so now I can do this." Rebreathers have raised the bar of what is possible in diving. But just because the tool is capable of doing the job, it may not mean that you are experi-

CHAPTER 11 - MYTHS AND MISCONCEPTIONS

enced enough to take it on. Even if you survive one particular dive, you may not survive the next one if your attention to safety is too relaxed. Our ever-quickening technical world offers many opportunities, but those opportunities carry definable risk that you and your family must assume if you choose to use rebreathers.

Myth: I'll be able to sense a CO_2 problem before it gets to be too late.

Divers under a high workload are already breathing hard. This can mask the onset of carbon dioxide problems that bring on dypsnea (shortness of breath). Carbon dioxide poisoning will not be found in an autopsy, but several accidents that have been recorded on helmet cameras have demonstrated very high respiratory rates and difficulty breathing. Carbon dioxide poisoning is a slippery slope. Once you start down that slope, it may be difficult to recover.

If you suspect a carbon dioxide issue, immediately bailout to open circuit and carry adequate gas supplies to return to the surface with a generous margin for safety. Don't expect clear symptoms to warn you of a CO_2 issue.

Myth: If I experience a CO_2 breakthrough I can switch to SCR mode to save gas.

Using the rebreather in SCR mode still exposes the diver to elevated levels of carbon dioxide. The cumulative exposure to high CO_2 will eventually lead to unconsciousness and drowning. The safest abort plan for carbon dioxide breakthrough is by using open circuit gas with a well-tuned regulator designed and maintained for the depths and gases you are diving.

Myth: I'm breathing therefore I am fine.

In open circuit diving, emergencies are often accompanied by a loud sound and gas loss. In rebreather diving, you are manipulating your life support environment. Problems can be silent and deadly if you are not paying attention to your displays. Problems tend to develop slowly on a rebreather, giving you time to solve issues, yet they can be more easily overlooked. A DAN study noted, "The inability of a diver to recognize when the breathing gas is unsafe (due to hypoxia, hyperoxia or hypercarbia) makes self-rescue unlikely." The bottom line- you are not likely to recognize a symptom before you are either incapacitated or unconscious.

CHAPTER 11 - MYTHS AND MISCONCEPTIONS

Myth: A BOV is a life saver.

There is a question of convenience versus complexity when it comes to BOVs. Some of the pros for BOVs include: easy switching to open circuit and never having to remove loop from your mouth. A diver experiencing dypsnea may have a very difficult time switching to OC. Respirations are rapid and the diver may fear they will drown. They clamp down hard on whatever is in their mouth. However, several cons include: increased drag and jaw fatigue in high flow and on scooters, must be plumbed to adequate gas supply (not onboard gas), new failure points, may not be "Class A" rated for depth, may not be usable with hypoxic diluent in shallow water and may introduce dead spaces.

Don't get me wrong. In recreational depths I have no issue with BOVs, but when I see the problems associated with extra hose routing, connections on some technical rebreathers, they may come in to question. Most importantly though, if your BOV is not routed to an adequate gas supply, then you still have to make a switch offboard, negating the best benefit. I suggest considering your particular diving goals and evaluating their benefits for yourself.

There may be a way to breathe your CCR in an "open loop" fashion. In other words, if the diluent addition valve makes gas flow into the loop close to the inhalation side of the mouthpiece, you will be able to breathe open circuit by injecting diluent and exhaling each breath while still in a closed circuit position on the DSV.

Myth: BOVs are great for taking a sanity breath.

BOVs do indeed allow for a quick switch to open circuit, however a "sanity breath" will not restore your full mental acuity if you are drifting into hypoxia. It will take considerable time to restore the balance to your body and reinstate your ability to think clearly. A sanity break of many minutes may be required.

Myth: My scrubber is good for four hours independent of depth.

Canisters "roll-off" the deeper you go. A scrubber that is stated as capable of a 4-hour runtime is likely quoted for 130 feet/40 meters depth limit and

Use manufacturer's recommendations and properly validated data to determine the duration of your scrubber. Make sure you have validated data for deeper dive applications.

CHAPTER 11 - MYTHS AND MISCONCEPTIONS

air diluent. Some rebreathers will only offer 25% of that time at 330 feet/100 meters. You will regain scrubber capacity as you ascend through decompression, but many profiles dived today exceed the tested limits of the unit. You might get away with it on a relaxed warm water dive, but during a high exertion rescue at the end of your bottom time, you might overbreathe your canister.

Myth: Five pounds of scrubber gives me five hours… in any rebreather.

Use a paper, electronic or automated checklist. You will get distracted during preparation. A checklist ensures you have not missed a step when you get back to work.

Depth obviously changes scrubber duration as mentioned above, but rebreather design can also affect duration. Five pounds in one rebreather may get far less duration in another due to design features that affect dwell time, loop temperature and other factors.

Myth: Doing my checklist from memory is okay.

Checklists are cool, but they need to be on paper or in a digital format. We get distracted during rebreather preparations. Checklists ensure you can pick up where you left off. We simply don't find many dead divers with checklists.

The manufacturer of your rebreather has spent years in testing and validation to get it ready for distribution. As a new rebreather diver you should carefully read and heed their warnings. Use the rebreather as they intended it to be used.

Myth: I can store my oxygen sensors in a ziplock bag of nitrogen or in my fridge to extend their life.

According to VR Technology's Kevin Gurr, oxygen sensors should be changed every 12 to 18 calendar months once opened. He also advises not to use sensors that have been stored for over two years. Finally, he advises that they should not be stored in a sealed bag.[10] Most sensor manufacturers back up these recommendations.

Myth: I can use medical sorb in my rebreather to save money.

Rebreathers are tested and validated using particular brands and grades of absorbent material. You should only use the brands recommended by the manu-

CHAPTER 11 - MYTHS AND MISCONCEPTIONS

facturer. Medical sorb has a different moisture content than diving sorb. It will not react in a predictable way inside a moist environment of exotic gases that are completely different from an anesthesia machine.

Myth: Any brand of oxygen sensor is okay to use.

Just like absorbent material, particular brands of sensors are approved for particular units. Months and sometimes years of testing by the manufacturer are needed to refine the calculations used to account for the environmental conditions inside a particular rebreather. The software interface in a rebreather has to be designed to provide the correct PO_2 on your display in a linear relation to the output of the sensors themselves. The electronics also need to be temperature compensated over the range of expected ambient temperatures. A sensor that has not been tested and validated in a particular unit may not behave in a predictable way. Use only sensor brands that are approved by the manufacturer.

As research proceeds, the knowledge base of our sport will expand. Diligent divers will have to be discerning consumers of information. By sharing our successes and failures, all divers can contribute to a better understanding of this technology and our interface with it.

[10] Gurr, Kevin - "Oxygen sensors for use in rebreathers." 2013. Accessible at: RebreatherPro.com.

Jake Rehacek using his sidemount rebreather at Devil's Eye Spring.

12

Culture of Rebreather Diving

Shifting the Culture of Rebreather Diving to Reduce Accidents and Incidents: Five Golden Rules

Rebreathers have been connected to approximately 20 deaths per year in the sport diving community.[11] Significant evidence supports that these fatalities are often tied to failures of the human machine interface (HMI) as well as risky choices and behaviors. This chapter aims to suggest a cultural shift in rebreather diving that minimizes and prevents future deaths. This shift consists of adopting five basic rules in your personal diving as well as insistence that your buddies follow the same responsible guidelines. The rules fall into the following categories: training/currency, checklists, pre-breathe, decision to dive and aborting dives. Whether you are new to rebreathers or an experienced CCR diver who feels very comfortable with them, these rules could save your life.

Rule 1. Recognize and prevent complacency in yourself and others around you.

Accidents are frequently labelled as "pilot error," so it behooves us to examine that nature of pilot error. Technical diving, and specifically rebreather diving, is a continual learning process. If we closely examine how we learn, we can better prepare for the pitfalls associated with each stage of the learning process.

When you operate with well understood and consistent practices, then everyone follows the same rules. Nobody should be exempt from safety practices and even the newest member of a team should feel comfortable questioning the actions of everyone, including the team leader. Numerous incidents have occurred when a team leader was not questioned on their actions or behaviors that subsequently led to an accident.

CHAPTER 12 - CULTURE OF REBREATHER DIVING

For example, an internationally recognized climber threads the rope through her harness on an easy climb. She is temporarily distracted by someone with a question and while answering, she stops to tie her shoes. She makes her climb and when she leans back to rappel, she falls 72 feet, narrowly escaping her death when cushioned by tree branches. In her case more training would not have helped. Experience actually contributed to her accident. She tied off when she was supposed to routinely tie off- but rather than her harness, it was her shoes.

Earlier in this text, I wrote about the Conscious Competence model, which describes the steps involved in the process of learning any new skill. This model is particularly applicable to rebreather diving.[12]

I have often thought that new rebreather divers with roughly 50 to 100 hours after their initial training, may be at the greatest risk in their diving careers, especially if nothing has scared them along the way. The human brain is exquisitely tuned to detect novelty but when everything becomes routine, we tend to stop paying attention. Yet, according to Dr. Jeffrey Schwartz at the UCLA School of Medicine, humans can literally alter the anatomy of the brain by the demands required of it.[13] He noted that the hippocampus of cab drivers grew larger as they learned the layout of a new city. The hippocampus is responsible for map making and he noticed profound neural changes that occurred within a few days. In rebreather diving, novelty helps us learn and expand our hippocampus, but we need to guard against complacency when diving becomes routine.

If you have come to rebreathers with a lot of open water experience, you should prepare yourself to step back for a while and gain hours until your intuition and skills are finely tuned on the rebreather. Only then should you consider moving back into the challenging environments and specialty applications that you were involved in before.

When a rebreather diver experiences a serious gear malfunction, it often frightens the diver back to the previous level of learning where they become conscious drivers of their unit again. A long absence from diving will also result in the diver stepping backwards in the model until they catch up with their skills and practice.

CHAPTER 12 - CULTURE OF REBREATHER DIVING

The climber who experienced the fall was attempting to use a behavioral script to prepare her harness. Using her memory, she conducted a series of steps she had done repeatedly without incident. At the moment when her mental model indicated that she should tie her harness, she was distracted and instead tied her shoe. This action likely satisfied her behavioral script and she moved on to the next phase of her climb. Her mental checklist had become routine.

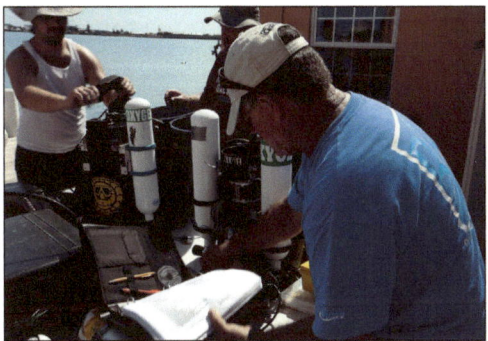

You might be tired of me pushing checklists, but the bottom line is that we rarely find checklists at fatal accident sites.

To avoid the pitfalls of complacency, proper procedures and a commitment to pre-dive checklists and pre-breathe sequences is critical. A diver who carefully reviews his personal preparedness as well as equipment readiness will be better poised to deal with the issues he may encounter ahead.

Rule 2. Always use a written or electronic checklist to prepare your rebreather.

Many fatalities could have been prevented by using a paper or onscreen checklist. Distractions and lapses in memory are human factors that we need to account for in all diving activities. It is incumbent on instructors, mentors and prominent divers to change the culture of rebreather diving to something equivalent to flight training. Checklists have to be viewed as "cool" to be popular. When failing to use a checklist is frowned upon at the grassroots level, then they will be more widely utilized. Role models and thought leaders need to be conspicuous in using checklists and insistent that their diving partners do the same.

Rule 3. Always conduct a five-minute pre-breathe in a safe, place with your nose blocked.

A five-minute pre-breathe should become the norm throughout our sport. Conduct this activity in a seated and safe location so you can observe your handsets, listen for your solenoid and evaluate your physical condition prior to reaching a place of danger where you could fall and injure yourself or drown. Initial pre-breathe sequences should never be conducted while walking to an entry area, putting on fins or floating in the water.

197

CHAPTER 12 - CULTURE OF REBREATHER DIVING

Never Got Wet

I was sitting in the Mission Control trailer when I heard someone yelling outside at the dive site. "911!" I heard, echoing through the trees.

I ran outside to find a CCR diver face down in the water. Three of us dragged the victim ashore and made every heroic to save the man. An hour of CPR and advanced cardiac life support did not bring him around. He was dead. Upon downloading his rebreather, we learned that he had not even been diving yet. While standing in shallow water, he was breathing on the loop and putting his fins on. A missed step in his pre-dive and he was slowly depleting the oxygen in the loop. Distracted while putting on fins, he passed out standing up then fell into the water and drowned.

Rule 4. Do not dive if your rebreather has not completely passed all pre-dive checks and tests.

Significant numbers of accidents and fatalities are attributed to high risk behaviors such as beginning a dive with a known technical fault like a single sensor failure. These faults may not be detected until a pre-breathe is attempted, so if it is still critical to do the dive, then it is incumbent on the diver to have alternative equipment available to make the dive safely.

Rule 5. Abort your dive in the safest possible mode.

In the early days of rebreather training, we encouraged divers to find a way to stay on the loop if possible. Culturally, this flawed practice may have led to incidents where divers felt it reflected poorly on them when they bailed to open circuit. In most cases today, a properly equipped CCR diver has sufficient personal gas to execute an open circuit bailout. If you have sufficient gas for abort, open circuit is the safest option and definitely preferred when diving alone.

If in doubt, bail out. There will always be another day for diving.

CHAPTER 12 - CULTURE OF REBREATHER DIVING

Whether the abort takes place on the boat deck or during a dive, the community needs to be encouraged to follow safe role model behaviors. Refuse to dive with those that do not abide by essential safety measures. Refuse to dive with those that take unnecessary risks for themselves and their team.

Dr. Andrew Fock's research has revealed that most fatalities are attributed to diver choices and behaviors rather than any particular model or style of rebreather.[14] Given that revelation, we have a unique opportunity to grow the market in a safer way by encouraging and applauding safe diving practices and focused attention to procedures.

I'm Snorkeling Today

On a dive in the Dominican Republic, I experienced an unusual failure on my rebreather. The unit completely turned off and then reset itself in 70 feet of water. When it reset, it pushed a large slug of oxygen into the loop. Startled, I shut down the oxygen valve momentarily to stop the flow of oxygen and aborted the dive. I assumed that I had some sort of a battery failure or intermittent power issue. That night I rigorously tested the unit and determined that everything was working to specifications with a new set of batteries. I prepped it for the next day's diving and loaded the boat. The pre-dive passed and pre-breathe was uneventful, but as I stood to walk to the back of the boat, the rig turned itself off and on again adding a slug of oxygen as it powered up. I sighed, sat down and told everyone onboard that I would not be shooting photos. I was going to snorkel instead. A volley of ideas were tossed my way including a suggestion to run my dive in manual control. I made it clear to my partners that I don't get in the water with gear that is already broken. It is not lost on me today, that one of the people encouraging me to dive that day has since died on his rebreather.

Rebreather Code of Conduct

As prudent divers, we have responsibilities to our loved ones and ourselves.

- ✓ You must accept personal responsibility for your actions and carefully manage your own risk assessment.
- ✓ You must dive within the limits of your training and experience.
- ✓ You must be aware that complacency creeps in to the practices of experienced divers and must be vigilant with checklists and pre- and post-dive safety procedures.
- ✓ Recognizing that rebreather diving requires alertness, you should carefully assess dive conditions and your own physical and mental preparedness prior to every dive.
- ✓ You must be familiar with your equipment and safe operating procedures and recognize the need for well-practiced and current skills.

CHAPTER 12 - CULTURE OF REBREATHER DIVING

- ✓ You should recognize your part in an evolving sport by staying current with developments and emerging knowledge within the industry.
- ✓ You must personally analyze your own gas as well as personally prepare your own rebreather, recognizing your responsibility for its safe operation.
- ✓ You must carry adequate bailout gas to help you recover from a catastrophic loop failure that happens at the worst possible moment.
- ✓ You should maintain good health and specialty diving accident insurance, such as DAN Insurance.
- ✓ You should share your motivations for rebreather diving with your family and have frank discussions about risk versus reward, allowing them to participate in the risk assessment process.
- ✓ You must service and maintain your equipment within manufacturer's guidelines.

Consensus Statements of Rebreather Forum 3.0[15]

The Rebreather Forum 3 (RF3), held in May 2012, in Orlando, Florida is the most important rebreather safety conference ever assembled. RF3 addressed the major issues surrounding rebreather technology and its application in commercial, media, military, scientific, sport and technical diving. Specialists, experts, experienced practitioners, manufacturers, and ancillary companies representing a variety of communities discussed technology and shared information. The program included dedicated sessions covering topics such as medicine and physiology, business and operations, CCR familiarization, training, design and testing, and incident analysis. At the end of the forum, the final session gathered all participants to determine consensus statements that reflected the state of the sport today and highlighted areas in need of research and attention.

RF3.0 brought together rebreather divers and experts from all over the world for seminars, workshops, pool experiences and consensus forums.

The RF3 consensus statements were written by Dr. Simon Mitchell and agreed upon by attendees present in the final session of the conference.

Checklists

The forum acknowledged the overwhelming evidence demonstrating the efficacy of checklists in preventing errors in parallel fields that share similar technical complexity. Two recommendations regarding checklists were consequently agreed:

CHAPTER 12 - CULTURE OF REBREATHER DIVING

Checklists 1:

The forum recommends that rebreather manufacturers produce carefully designed checklists, which may be written and/or electronic, for use in the pre-dive preparation (unit assembly and immediate pre-dive) and post-dive management of their rebreathers.

Written checklists should be provided in a weatherproof or waterproof form.

The current version of these checklists annotated with the most recent revision date should be published on the manufacturer's website.

Checklists 2:

The forum recommends that training agencies and their instructors embrace the crucial leadership role in fostering a safety culture in which the use of checklists by rebreather divers becomes second nature.

Training and Operations

Training and Operations 1:

The forum applauds and endorses the release of pooled data describing numbers of rebreather certifications by training agencies, and encourages other agencies to join ANDI, IANTD, and TDI in this initiative.

Training and Operations 2:

The forum endorses the concept of making minimum rebreather training standards available in the public arena [accessible to the public].

Training and Operations 3:

The forum endorses the concept of a [diving and knowledge] currency requirement for rebreather instructors. We recommend that training agencies give consideration to currency standards in respect of diving activity, class numbers [number of classes taught] and unit specificity for their instructors.

CHAPTER 12 - CULTURE OF REBREATHER DIVING

Training and Operations 4:

The forum recognizes and endorses the industry and training agency initiative to characterize "recreational" and "technical" streams of sport rebreather diver training. These groups will have different operational, training and equipment needs.

Accident Investigation

Accident Investigation 1:

The forum recommends that training agencies provide rebreather divers with a simple list of instructions that will mitigate common errors in evidence preservation after a serious incident or rebreather fatality.

These instructions will be developed under the auspices of the Undersea and Hyperbaric Medical Society Diving Committee in consultation with the relevant RF3 presenters.

Accident Investigation 2:

The forum endorses the concept of a widely notified centralized "on-call" consultation service to help investigators in avoiding errors or omissions in the early stages of a rebreather accident investigation, and to facilitate referral to expert investigative services.

Accident Investigation 3:

The forum recommends that in investigating a rebreather fatality, the principal accident investigator invite the manufacturer of the incident rebreather (or other relevant equipment) to assist with its evaluation (including the crucial task of data download) as early as is practicable.

Very clear lines were drawn to differentiate between recreational and technical sport diving. This will help clarify skills training, support requirements and design guidelines for the two classes of rebreathers.

CHAPTER 12 - CULTURE OF REBREATHER DIVING

Accident Investigation 4:

The forum endorses the DAN worldwide initiative to provide a means of on-line incident reporting with subsequent analysis and publication of incident root causes.

Design and Testing

Design and Testing 1:

The forum recommends that all rebreathers incorporate data logging systems which record functional parameters relevant to the particular unit and dive data, and which allow download of these data. Diagnostic reconstruction of dives with as many relevant parameters as possible is the goal of this initiative.

Note: An ideal goal would be to incorporate redundancy in data logging systems, and as much as practical, to standardize the data to be collected.

RF3 recommended that data-loggers be incorporated into rebreathers. This will create the biggest challenge for the manual class of rebreathers.

Design and Testing 2:

The forum endorses the need for third party pre-market testing to establish that rebreathers are fit for purpose. Results of a uniform suite of practically important unmanned testing parameters such as canister duration, and work of breathing (qualified by clear statements of experimental parameters) should be reported publicly. Ideally, this testing should be to an internationally recognized standard.

Design and Testing 3:

The forum acknowledges recent survey data[16] indicating a poor understanding of rebreather operational limits in relation to depth and carbon dioxide scrubber duration among trained users, and therefore recommends:

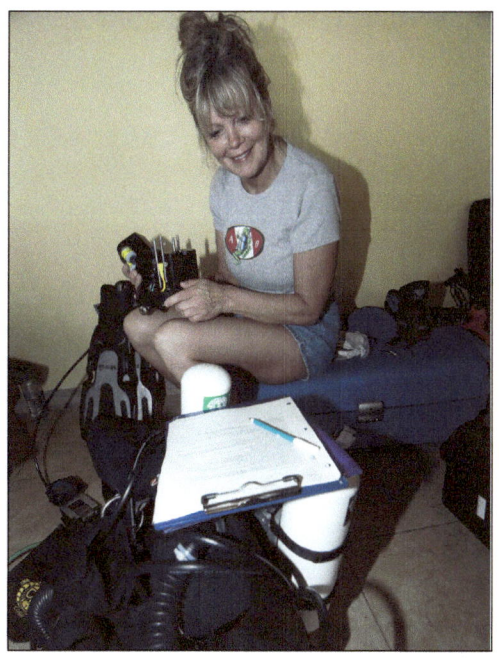

RF3 recommended that certain test data using standardized methods be made openly available to the public.

1. that training organizations emphasize these parameters in training courses.

203

CHAPTER 12 - CULTURE OF REBREATHER DIVING

2. that manufacturers display these parameters in places of prominence in device documentation and on websites.

Design and Testing 4:

The forum strongly endorses industry initiatives to improve oxygen measurement technologies, and advocates consideration of potentially beneficial emerging strategies such as dynamic validation of cell readings and alternatives to galvanic fuel cells.

Design and Testing 5:

The forum identifies as a research question the issue of whether a mouthpiece retaining strap would provide protection of the airway in an unconscious rebreather diver.

Design and Testing 6:

The forum identifies as a research question the efficacy of full face masks for use with sport rebreathers.

RF3 was undoubtedly the most important gathering of rebreather divers and experts ever assembled. Conference recommendations may take time to implement, but it should be noted that the majority of recommendations directly address the culture of rebreather diving and how we utilize technology. They are important directives that role model divers should actively embrace.

---

[11] Fock, Andrew. RF3.0 Proceedings.
[12] This Learning Stages model was developed by former GTI employee, Noel Burch, http://www.gordontraining.com/free-workplace-articles/learning-a-new-skill-is-easier-said-than-done/
[13] Rewire your Brain, Aalto University Executive Education, Feb 03, 2012, http://www.slideshare.net/AaltoEE/rewire-your-brain-11400486
[14] Rewire your Brain, Aalto University Executive Education, Feb 03, 2012, http://www.slideshare.net/AaltoEE/rewire-your-brain-11400486
[15] Presentations and Consensus Statements are available as PDF files and video download at www.rf30.org.
[16] http://rubicon-foundation.org/Projects/rebreather_canister/

Rebreather Divers Deploy a Technological Arsenal to Explore a Twilight Zone of Ice Age Reefs

We had been planning this dive over our entire careers. Our mission: Descend more 200 feet in open water, find the peak of a majestic volcanic seamount, then drop to below 450 feet to search for signs of ancient sea levels on what was once a small Atlantic atoll. Our dive would more than double the depth of previous underwater expeditions into the secretive Bermuda Deep.

Biologist Dr. Tom Iliffe and I scanned each other for bubbles and positioned our open circuit bailout regulators within close reach. We nodded, and took one last look at the support team on-board the *Pourquois Pas* – *"Why Not?"* – the name etched on the transom. I chuckled to myself. Iliffe and I glanced at each other, dumped the gas from our wings and rocketed downward, racing to ensure every possible moment would be spent gathering data and documenting our work. I film his descent from above, a tiny figure disappearing into the void.

We reach the timeworn peak of the formidable Challenger Bank in under three minutes. A lonely lionfish hunkers in the clumping masses of coral where our hook jerks and bounces, dangling slightly, taut at its maximum reach. The orange surface marker is long out of sight, bobbing overhead on the waves marking our position. Moments like this make me acutely aware of my humanity, swimming awestruck in a place that has been out of reach until now.

Iliffe swiftly ties on his cave diving reel and flies out over the precipitous drop. It seems like an eternity before he lands on a tiny crag at 460 feet, with the endless wall sloping infinitely downward and out of sight.

He locks off the dive reel, positions his mesh bags and tools then intently goes to work. I train the camera on my usually reserved friend, preparing to document careful sampling. But Illiffe is like a Walmart shopper crashing the door on Black Friday. We have only minutes at this depth, and he's making the most of every breath. Hammer swinging and arms flailing, he grabs rock samples and delicate coral twigs. As though he were loading Noah's Ark, he bags sets of animals, ensuring he has at least two of everything new or interesting.

I alternate between photographing his feats and the delicate wall, covered with fragile purple hard corals and crusting fiery sponges, flaming in bursts of color and vibrance. Schools of curious jacks zip around us while huge crab-eating permit reflect the sultry light back towards us. In this previously unexplored twilight zone, there is no shortage of extraordinary life to observe.

Illiffe collected more than 50 species of plants and animals on those dives that today are yielding new discoveries for Bermuda, some previously unknown to biologists. Geologists are studying photographs and rock samples trying to piece together the changes in sea level over time. Examining deep cave structures and wave cut notches they can now determine when the sea level was at its lowest point. The first glimpse into Bermuda's twilight zone suggests that exploration and discovery are still in the very early stages. Future work will focus on determining how the unique life in Bermuda first populated remote island caves. Did the cave-adapted animals migrate upwards from deep ocean vents, swim through tiny spaces within the

Brian Kakuk and Paul Heinerth hang out on decompression. The support team from Triangle Diving looks on from above.

matrix of rock or arrive in some other way? Indeed, we know more about outer space than the pristine Bermuda, Argus and Challenger Banks. For those lucky enough to be working there, Bermuda Deep offers the chance to join the ranks of aquanauts on the edge of underwater discovery.

For further info on the project: http://oceanexplorer.noaa.gov/explorations/11bermuda/welcome.html

13 Travel

Tips and Warnings

Traveling with rebreathers can have its challenging moments. Fortunately, representatives from the Transportation Security Administration (TSA) in the U.S. attended Rebreather Forum 3.0 to expand their understanding of the technology. That said, there are still horror stories of problems faced by travelers with rebreathers. In one situation, the conversation between a TSA agent and a scientific diver went like this:

"Sir, you cannot take tanks in your luggage."

"But I have removed the valves."

"Sorry Sir, pressurized cylinders are not allowed on the plane."

"But they are not pressurized. They are just a chunk of metal."

"I'm sorry Sir, you will need to prove to me that they are not pressurized."

At this point the diver unpacked his gear, dug out the tank valves, placed them back in the tanks, applied a regulator and showed the TSA agent "zero" on the gauge when he opened the valves. The agent now satisfied, allowed my colleague to re-pack and travel with his tanks.

Unfortunately, not all agents have an education in SCUBA diving and their familiarity with diving gear may be minimal. I have developed a good rapport with my small local airport. I get the "oh, you again" treatment. We've been through the educational stuff together and I am always polite and ready to teach them about anything unfamiliar.

CHAPTER 13 - TRAVELING WITH REBREATHERS

There are a few things you can do to make traveling easier.

- If you can, avoid carrying sorb; it is generally cheaper to get it at your destination anyway. In Florida it is inexpensive and easy to get at roughly $125 per 45 pound/20 kilo keg. Many other destinations supply sorb and onboard cylinders for visiting divers. It is usually cheaper to rent tanks than pay extra baggage fees. If you are traveling somewhere without support, then you may have to carry sorb in your checked baggage.
- Download the MSDS (Material Safety Data Sheet)[17] from the Internet for the brand of sorb that you are carrying. Tape the MSDS to the keg and if you feel like being a bit creative, print a sticker on your computer that reads "Safe for Airline Transport" in large block letters and stick it to the keg.

- I like to put a formal letter in each case of luggage that begins: "Dear TSA Agent..." Essentially, you want to advise them that this is a diving apparatus known as a rebreather and that it is delicate life support equipment. Put your cell phone and seat assignment on the letter and urge them to call if they have questions. Thank them for their diligence in securing your life support equipment. Put a small bundle of replacement zip ties inside the luggage so they can re-secure the zipper pulls or case latches. If you have a title, logo or company letterhead that looks impressive, put that on the letter. If you have photograph of your smiling face in that diving equipment, print that on the letter. If this is the trip of a lifetime, tell them. Making their job easy, fun and light will increase your chances of getting to the destination with all gear intact.
- Tanks must have the valve removed and must not be covered with tape or plugs. Place tanks in a large plastic bag so that the screener can look into the tank without removing it from the bag. Take a small VIP light to your destination to check tanks before replacing the valve. Note: your oxygen clean status is now invalid.
- Cover the oxygen label on your tank with duct tape. If screeners see the word oxygen, they usually go into full emergency mode recalling the oxygen tank that brought down the plane over the Florida Everglades. Stick a plastic flower in your tank, then when they tell you tanks are prohibited, tell them it is a big steel flower vase.
- Spare lithium batteries may not be carried checked in baggage. They may be installed in the rebreather and two spare lithium batteries may be hand carried.

CHAPTER 13 - TRAVELING WITH REBREATHERS

Preventing lithium batteries from short circuiting is very important to impeding the likelihood of overheating and fire. Always keep lithium batteries isolated from metal objects (e.g. jewelry, keys) or other conductive materials by enclosing each one separately and insulating terminals with a non-conductive material (e.g. electrical tape). Pack them so they cannot shift during transport. Physically damaged or dropped cells can become volatile. A lithium battery inside equipment is protected from short circuiting because it is secured in the actual device and cannot move around during transport.

- Make sure no switches or power buttons can be accidentally turned on during transport.
- Regulations are always fluctuating, so keep up with changes at the TSA website.[18]

How much lithium can you take on an airplane?

Equivalent Lithium Content (ELC). ELC is a measure by which lithium ion batteries are classified.

8 grams of equivalent lithium content are equal to about 100 watt-hours.

25 grams of equivalent lithium content are equal to about 300 watt-hours.

Batteries greater than 25 grams not permitted.

Watt-hour (Wh) Rating for Lithium

The Wh indicates the amount of energy contained in a lithium battery. The UN Recommendations on the Transport of Dangerous Goods, Model Regulations regulate Li-ion batteries based on their Wh rating.

How to calculate the Wh rating:

The Wh rating must appear on the battery case if it was made on or after January 1, 2009. If it is not there, you can calculate the Wh rating by using one of these formulas:

If you know the nominal voltage (V) and the capacity in ampere-hours (Ah), then Wh = (V) x (Ah); or

If you know the nominal voltage (V) and the capacity in milliampere-hours (mAh), then Wh = (V) x (mAh ÷ 1000).

If you are still not sure of your lithium battery's Wh rating, contact its manufacturer.

How to calculate the lithium content:

You can calculate the lithium content, in grams (g), of a lithium metal cell:
If you know the battery's capacity in ampere-hours (Ah), then Grams (g) lithium metal = (Ah) x 0.3; or
If you know the capacity in milliampere-hours (mAh), then Grams (g) lithium metal = (mAh ÷ 1000) x 0.3.
To calculate the lithium content of the battery, simply multiply the grams (g) of lithium metal by the number of cells in the battery.

CHAPTER 13 - TRAVELING WITH REBREATHERS

- Pack your rebreather in a bag that falls within the weight guidelines of 50 pounds or 23 kilos. Separate things into multiple cases if necessary. Overweight cases get more scrutiny than gear that fits within standards. Cases get more scrutiny than bags. Put a plastic bin inside a duffle or roller bag to protect your investment and remain low key.
- Consider renting rebreather tanks on arrival at your destination. Each 19 cft/3 liter tank weighs a hefty ten pounds or 4 kilos in baggage. 300 bar DIN valves may not be filled in many European destinations. Regulations restrict some dive shops from filling these valves even at lower pressures. They may not even have a 300 bar fill adapter available. It may be easier to use 200 bar DIN valves. Some countries require hydrostatic tests at a greater frequency than once every five years. (Some require tests be conducted every two years). Your tanks will have to meet local regulations for filling and renting them locally will save headaches. I recall landing in Lanzarote with our rebreathers. We rented dozens of bailout tanks and K-bottles of gas, but had to get all of our rebreather tanks hydrostatically tested by the local Spanish authority. They simply did not want to accept our DOT testing from the U.S.
- Pack neatly since disheveled cases get more attention.
- Photograph the interior and exterior of your packed bags and make an inventory of contents. In the event that your bag is lost or parts are damaged, you have vital proof for a claim with the airlines or tour insurance carrier.
- Consider extra insurance for your equipment. DAN has a good partner for equipment insurance.[19]
- Bring adequate spare parts and tools to repair any damage on arrival, but do not carry large tools in carry-on baggage. Deadly attacks with Allen keys and wrenches must be on the rise, because there are strict limitations on the size of tools that may be hand-carried on the aircraft.
- Last but not least… drive if you possibly can or pre-ship if you can afford it.

Mask, fins, rebreather and you are ready to go, right? There are many small items that you might not think about that can ruin a trip in their absence.

Inquire about the local power supply. As an example, Mexico has the same current as the U.S., but many wall outlets do not have grounding plugs. You will need to bring along or purchase "cheaters" that convert your

CHAPTER 13 - TRAVELING WITH REBREATHERS

three-prong plugs to two-prong. Power strips are worth their weight in gold since wall outlets may be few and far between. In Europe, recessed wall plugs may not fit the fancy converter you bought at the airport. Those converters also often lack a hole for the grounding plug, making them impossible to use without the "cheaters" mentioned above. Many chargers are already rated for 120 and 240v power supplies. Do not waste converters on these devices as they are not needed.

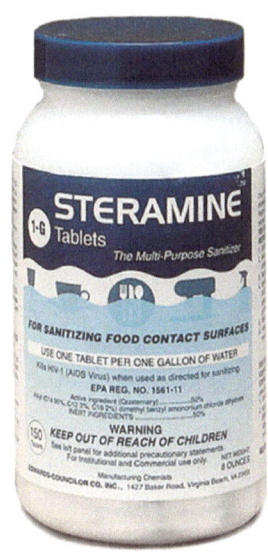

Disinfectant may be challenging to take to your destination. Virkon or Steramine tablets are very convenient for travel since they are dry and will not spill in luggage and won't take up precious weight allowance. Betadine is usually easy to find in foreign pharmacies (though some people are allergic to Betadine). Dilute it in the bathtub to soak your breathing loop. If you are camping on an expedition, use one of your shipping cases as a bath. If you cannot find Betadine, Listerine will get your through a trip with reasonable cleanliness, but it does not kill all bacteria. Use proper disinfectants as soon as available.

If you are traveling in a small group, look into private charter flights. Your baggage allowance may be greater and screening much easier since you will travel from a smaller airport or FBO for private planes. I have flown to and from film shoots on charter planes with a scooter between my legs and a half ton of gear piled in the extra seats around us.

You can never give enough credit to duct tape. It can repair an awful lot of damage in a difficult situation. I always carry some on trips, but rather than taking a heavy roll, I just wind a few yards around a business card for a small supply. A product called "Rescue Tape" is also fantastic. It is made of clear silicone self-binding tape. It does not have a sticky backing, but instead, bonds to itself in mere moments. It is extremely useful.

Strong zip ties and a Leatherman-type multi-use tool are extremely useful.

Clear plastic Ziplock bags allow you to neatly separate and pack gear so that it does not get lost during inspections. If you simply roll your DSV in your pajamas, the screener is likely to pick up the garment and send the sensitive device crashing to the floor in pieces. Slide delicate items into dive boots and fin pockets. Tape them in place or put them in Ziplock bags if necessary.

CHAPTER 13 - TRAVELING WITH REBREATHERS

Always make a photocopy of your critical identification like your passport and carry it separately in case of loss. Consulates can easily assist if you have back-up documents. Also hand-carry documentation about your rebreather and MSDS sheets for sorb or any other questionable items. Be prepared to share this information with screeners.

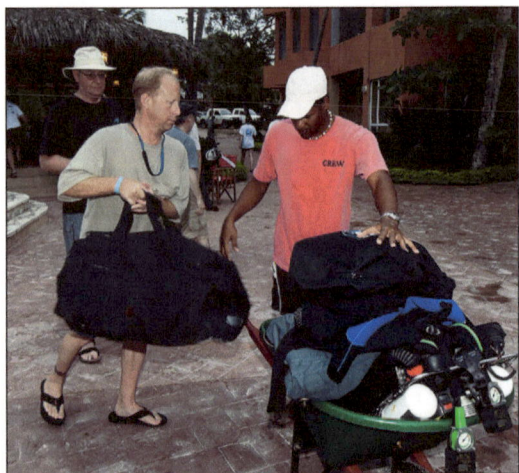

Traveling by Boat or Car

Your dive destination may be a lot closer to home, but there are still a few special tips to consider.

- Bring extra bungee cords for boat travel. Many boats are specifically constructed for single tank divers. The benches may be awkward for a rebreather diver. Bring your own method of securing the rebreather to the bench or the floor.
- Consider the orientation of your rebreather scrubber. If your rebreather is lying on its back, vibrating on a moving boat or in a car for hours, will your particular canister design be subjected to drastic settling? Some canister designs are equipped with springs that help to resolve settling issues, but if the orientation of the rebreather in relation to the settling forces is incorrect, it could result in channeling of material.
- Fully assemble and check your rebreather before leaving the dock. You don't want to be packing sorb on a bouncing boat and also don't want to capture diesel fumes within the loop.
- Beware of extreme heat. If you gear is sitting sealed in a hot car for a long period of time, you can damage the oxygen sensors. Diving a hot loop is also not a good idea.
- Beware of extreme cold. If your packed rebreather sits in a freezing car overnight, the moisture in the sorb can freeze causing damage to the canister and dusting in the material.

The anticipation of a trip is part of the excitement. If you apply the same enthusiasm to your gear preparation and packing, then your travel plans should be like smooth sailing.

[17] http://www.divegearexpress.com/rebreathers/absorbent.shtml
[18] http://www.tsa.gov
[19] http://www.diversalertnetwork.org/insurance/index.asp

14 CONCLUSION

In Conclusion

Now you have had a chance to consider the basics of rebreather diving and are ready to make a big decision in your life. It took me a lot of scrape together the funds to start my journey, but I have remained committed to the adventure of diving without bubbles. To say that I never looked back would be untrue. As I mentioned before, I have lost good friends to bad decisions. Each one of those people left a gaping hole in my heart. They've been present in my mind as I have penned each and every page of this book.

On the other hand, rebreathers have offered me incredible opportunities to go to places never seen by mankind. They have allowed me to interact with wildlife in a way that would be impossible on traditional SCUBA. They have become a remarkable tool in my dive locker.

As I have gained experience, I have become more discerning about who I dive with. I have become more vocal about safety issues and more diligent in my own dive preparation. I realize that I have much more to learn in this sport. I was not always patient. I was not always as careful as I am now. Sometimes I was even lucky to get home safely.

If you choose to take on rebreather diving, enjoy the opportunities. Be committed to safe diving practices and help others around you understand the risks and rewards that they face.

Use the best tools you have available to reveal Earth's secrets and then share your experiences with the world. It is good to be excited, stimulated and improved by the work of others. I hope that you can remember to share your pursuits and failures, your accomplishments and injuries so that we may all learn to be better, safer divers.

GLOSSARY

A

absolute pressure – the total pressure imposed by the depth of water plus the atmospheric pressure at the surface.

absorbent pads – absorbent material placed in a breathing loop; used to soak up moisture caused by condensation and metabolism.

accumulator – a small chamber that provides a collection vessel to ensure proper gas flow of oxygen to a solenoid valve.

active-addition – a rebreather gas-addition system that actively injects gas into the breathing loop (such as a constant-mass flow valve in certain kinds of semi-closed rebreathers).

ADV – see automatic diluent valve.

atmospheres absolute (ATA) – the absolute pressure as measured in atmospheres.

ATM – see atmosphere.

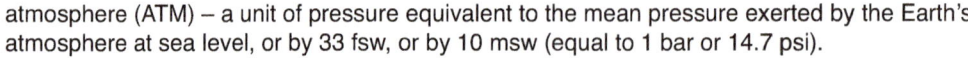

A = Active SCR - The KISS Classic is an Active SCR.

atmosphere (ATM) – a unit of pressure equivalent to the mean pressure exerted by the Earth's atmosphere at sea level, or by 33 fsw, or by 10 msw (equal to 1 bar or 14.7 psi).

automatic diluent valve (ADV) – a mechanically-activated valve that adds diluent gas when increasing pressure associated with descent or lowered volume triggers the device.

axial scrubber – a type of carbon dioxide absorbent canister design. In this design, the gas flows through the canister in a linear fashion from one end of the canister to the other.

B

backplate – a plate made of stainless steel, aluminum or ABS plastic which attaches to a rebreather and allows for the use of a webbed or soft harness system.

bailout – a failure requiring a dive to be terminated, usually using open circuit gas.

bailout gas – tanks carried by the diver to allow for escape from a serious situation, often conducted with open circuit technique.

bailout valve (BOV) – an open-circuit regulator built into the mouthpiece assembly that allows a diver to switch from closed-circuit mode to open circuit without removing the mouthpiece from their mouth. When the loop is closed, the BOV activates, supplying open-circuit gas directly from the onboard diluent tank (CCR) or supply gas cylinder (SCR).

bar – a unit measure of pressure, roughly equivalent to 1 ATM.

barotrauma – a pressure related injury.

GLOSSARY

B = Brian Kakuk (left) and Kenny Broad (right) - Retrieving a Lucayan skull in Andros, Bahamas during a National Geographic project.

BCD – see buoyancy control device.

bottom-out (counterlung) – a term used to refer to the situation when a rebreather counterlung becomes completely collapsed after a full inhalation.

boom scenario – an explosion or implosion of a hose or other component usually resulting in rapid gas loss or catastrophic loop failure.

BOV (bailout valve) – a mouthpiece block that allows a diver to switch from closed-circuit mode to open circuit without removing the mouthpiece from their mouth.

Boyle's Law – the volume occupied by a given number of gas molecules is inversely proportional the pressure of the gas.

breathing hose – large bore hoses in a rebreather breathing loop, through which the breathing gas travels.

breathing loop – the portion of a rebreather through which gas circulates, usually consisting of a mouthpiece, breathing hose(s), counterlungs, non-return valves and a carbon dioxide absorbent canister.

buddy lights – warning lights that indicates system status including life-threatening oxygen levels; usually monitored by the buddy diver.

buoyancy control device (BCD) – an inflatable bladder which allows a diver to precisely adjust buoyancy.

C

carbon dioxide (CO_2) – waste gas generated by the process of metabolism and exhaled by the diver into the breathing loop.

catastrophic loop failure – a complete failure of the breathing loop of a rebreather such that it cannot be recovered in closed-circuit mode; usually occurring from a ripping or tearing and subsequent flooding of a unit or a carbon dioxide emergency.

CCR – see closed-circuit rebreather.

central nervous system (CNS) – the human brain, spinal cord, and associated major neurological pathways that are critical for basic life-support processes, muscular and sensory systems.

cf (also cft) – cubic feet.

C = Christmas Island - Craig Challen (right) with John Dalla-Zuanna (left) - Diving Flex rebreathers on a National Geographic Project in the Indian Ocean.

check valve – a one-way, non-return valve that directs gas in one direction through the breathing loop.

215

GLOSSARY

closed-circuit rebreather (CCR) – a type of rebreather that usually includes some form of oxygen control system and generally only vents gas upon ascent.

CNS – see central nervous system.

CNS oxygen toxicity – a form of oxygen toxicity that affects the central nervous system; usually caused by exposure to breathing mixtures with an oxygen partial pressure in excess of 1.6 ATA. Symptoms may include visual disturbances, hearing anomalies, nausea, twitching, dizziness and severe convulsions as well as others.

C = Checklist - Use a checklist to prepare your rebreather.

CO_2 – see carbon dioxide.

CO_2 absorbent – a material that chemically binds with CO_2 molecules (Sodasorb, Drägersorb®, lithium hydroxide, Sofnolime®, Micropore ExtendAir, etc.).

CO_2 absorbent canister – a canister in the breathing loop containing CO_2 absorbent.

condensation – water that forms when water vapor cools and forms liquid droplets. In a rebreather, heat conduction through the breathing hoses and other components of the breathing loop lead to condensation. This process may be exacerbated by materials with greater heat conductivity and lessened with insulation of the breathing loop components.

constant-mass flow valve – a type of valve that allows a constant mass of gas molecules to flow at a fixed rate.

counterlung – a collapsible bag connected to a rebreather breathing loop, which expands as a diver exhales and collapses as a diver inhales.

cubic feet (cf) – a unit measure of volume, defined as the space occupied by a cube one foot on each side.

D

Dalton's Law - (also called Dalton's law of partial pressures) states that the total pressure exerted by the mixture of gases is equal to the sum of the partial pressures of individual gases. This empirical law was observed by John Dalton in 1801.

dcCCR – Diver Controlled Closed-Circuit Rebreather. A manually operated rebreather which requires the diver to monitor oxygen levels and manually inject oxygen as needed to maintain an appropriate setpoint. Also known as a Manual CCR.

D = Deco Support - Brett Gonzalez checks in on Brian Kakuk on a deep mission in Bermuda.

DCI – see decompression illness.

DCS – see decompression sickness.

GLOSSARY

decompression – the process of slowly ascending after a dive, while allowing excess dissolved gases in the diver's tissues to diffuse through the blood to the lungs and out with the exhaled breath.

decompression dive – any dive that requires staged decompression stops prior to ascending directly to the surface.

decompression sickness (DCI) – barotrauma including arterial gas embolism (AGE) and decompression sickness (DCS).

decompression sickness (DCS) – barotrauma seen especially in divers, caused by the formation of inert gas bubbles in the blood and tissues following a sudden drop in the surrounding pressure, as when ascending rapidly from a dive, and characterized by severe pains in the joints, skin irritation, paralysis and other symptoms.

demand regulator – a valve that delivers gas when the diver's inhalation reduces pressure.

diffusion – the process in which molecules move from a region of high concentration to a region of low concentration.

diluent – a cylinder in a closed-circuit rebreather that contains a supply of gas which is used to make up the substantial volume within the breathing loop; a mixture capable of diluting pure oxygen.

diluent purge valve/diluent addition valve – a manual valve used to add diluent gas to a breathing loop, usually through the counterlung or a gas block assembly.

DIVA (Display Integrated Vibrating Alarm) – an LED heads-up display module mounted close to the diver's mask, offering information about various states of the rebreather such as PO_2; this style includes a vibrating warning alarm when oxygen levels are unsafe.

downstream – a relative direction with respect to the flow of gas through the breathing loop of a rebreather; the direction of travel of the diver's exhaled gas.

downstream check-valve – a one-way, non-return valve that directs exhaled gas to flow from the mouthpiece to the exhalation counterlung. These mushroom-type valves prevent subsequent re-inhalation of used gas, and promote circulation of exhaled gas towards the carbon dioxide scrubber canister.

dynamic setpoint – also referred to as a floating setpoint, it is a setpoint that changes to optimize gas use, no stop time and other consumables and dive variables. The floating setpoint can be determined by an electronic system or modified manually by a diver using an mCCR.

E

EAD (Equivalent Air Depth) – a formula used to help approximate the decompression requirements of nitrox. The depth is expressed relative to the partial pressure of nitrogen in a normal breathing air.

EAN – see enriched air nitrox.

eCCR – an electronically controlled closed-circuit rebreather in which an electronics package is used to monitor oxygen levels, add oxygen as needed and warn the diver of develop-

E = Explorer End Cap - The spring loaded end cap has a visual "go/no go" button that can be seen through the clear housing.

GLOSSARY

F - Fill Station - Get your fills at a reputable fill source that analyzes their gas for quality on a regular basis.

ing problems through a series of audible, visual and/or tactile alarm systems.

Electronically monitored mSCR – a mechanical SCR with electronic monitoring. Electronics are used to inform the diver of PO_2 as well as provide warnings and status updates, however the gas control is manually controlled by the diver.

END (Equivalent Narcotic Depth) – a formula used as a way of estimating the narcotic effect of a breathing mixture such as heliox or trimix.

eSCR – an electronic semi-closed circuit rebreather where an electronics package monitors the PO_2 and adds gas to maintain a floating setpoint that optimizes gas use and compensates for changing levels of diver exertion.

enriched air nitrox (EAN) – a gas mixture consisting of nitrogen and oxygen; with more than 21 percent oxygen.

exhalation counterlung – the counterlung downstream of the diver's mouthpiece.

F

ffw – water depth as measured in feet of freshwater.

floating setpoint – (dynamic setpoint) a setpoint that changes to optimize gas use, no stop time and other consumables and dive variables. The floating setpoint can be determined by an electronic system or modified manually by a diver using an mCCR.

flush (as in flushing the loop) – replacing the gas within the breathing loop by injecting gas and venting bubbles around the edge of the mouthpiece or through a vent valve.

FHe – the fraction of helium in a gas mixture.

FN_2 – the fraction of nitrogen in a gas mixture.

FO_2 – the fraction of oxygen in a gas mixture.

fraction of gas – the percent of a particular gas in a gas mix.

fraction of inspired gas – the fraction of gas actually inspired, or breathed, by the diver.

fraction of inspired oxygen (FiO_2) – the fraction of oxygen inspired by the diver. In SCR operation, this figure is calculated using a formula that takes into account the diver's workload.

fsw – water depth as measured in feet of seawater.

G = Gauges - You need to be able to read your SPGs for your onboard tanks as well as your backup computers.

GLOSSARY

G

Gas narcosis – a form of mental incapacity experienced by people while breathing an elevated partial pressure of a gas.

H

harness – the straps and/or soft pack that secures the rebreather to the diver.

He – see helium.

heads-up display (HUD) – an LED display module mounted close to the diver's mask offering information about various conditions within rebreathers, such as PO_2.

heliox – a binary gas mixture consisting of helium and oxygen.

L - Loop - Safeguard your loop on the surface. Close the loop if it is out of your mouth.

helium (He) – an inert gas used as a component of breathing gas mixtures for deep dives because of its very low density and lack of narcotic potency.

Henry's Law – the amount of gas that will dissolve in a liquid is proportional to the partial pressure of the gas over the liquid.

HP Sodasorb – a brand of carbon dioxide absorbent material.

HUD – see heads-up display.

hydrophobic membrane – a special membrane that allows gas to flow through it, but serves as a barrier to water.

hyperbaric chamber – a rigid pressure vessel used in hyperbaric medicine. Such chambers can be run at absolute pressures up to nearly six atmospheres and may be used to treat divers suffering from decompression illness.

hyperbaric medicine – also known as hyperbaric oxygen therapy, is the medical use of oxygen at a higher than atmospheric pressure.

hypercapnia/hypercarbia – symptoms induced when the concentration of carbon dioxide in the inhaled breathing gas is too high, including shortness of breath, headaches, warmth, facial tingling and blackout.

hyperoxia – a concentration of oxygen in the breathing mixture that is not tolerated by the human body, generally occurring when the inspired PO_2 rises above about 1.6 ATA. Symptoms include visual and auditory disturbances, nausea, irritability, twitching, and dizziness; hyperoxia may result in convulsions and drowning without warning.

hypoxia – a concentration of oxygen in the breathing mixture that is insufficient to support human life, generally occurring when the inspired PO_2 drops below about 0.16 ATA.

I

inhalation counterlung – the counterlung upstream from the diver's mouthpiece block.

integrated open-circuit regulator – a second-stage, open-circuit regulator which is built-in to a mouthpiece block; also known as a bailout valve or BOV.

GLOSSARY

Inert Gas Narcosis – a form of mental incapacity experienced by people while breathing an elevated partial pressure of an inert gas.

L

LCD – Liquid Crystal Display.

LED – Light Emitting Diode; a small, low power light source used for warning lights on rebreathers.

LiOH – see lithium hydroxide.

lithium hydroxide (LiOH) – a type of CO_2 absorbent material.

loop vent valve – The adjustable overpressure-relief valve attached to the bottom of the exhalation counterlung, which allows excess gas and accumulated water in the breathing loop to be vented. Also known as an OPV.

M = Mouthpiece - Check your mouthpiece for holes or hidden tears that can cause a catastrophic loop failure.

M

manual bypass valve – a valve on a rebreather that allows the diver to manually inject gas into the breathing loop.

manual diluent addition valve – the valve on a rebreather that allows diluent gas to be manually injected into the breathing loop.

manual oxygen addition valve – the valve on a rebreather that allows oxygen to be manually injected into the breathing loop.

Maximum Operating Depth (MOD) - the maximum operating depth of a breathing gas before reaching a predetermined PO_2, usually 1.4 or higher. This depth is determined to safeguard the diver from oxygen toxicity.

mCCR – a manually operated closed circuit rebreather which requires the diver to monitor oxygen levels and manually inject oxygen as needed to maintain an appropriate setpoint. Also known as dcCCR or diver controlled CCR.

metabolism – the physiological process where nutrients are broken down to provide energy. This process involves the consumption of oxygen and the production of carbon dioxide.

mixed-gas rebreather – any kind of rebreather that contains a gas mixture other than pure oxygen in the breathing loop.

mouthpiece – the portion of a rebreather breathing loop through which the diver breathes. This usually includes a way to prevent water from entering the breathing loop and sometimes includes an integrated open circuit regulator (BOV).

MSW – water depth as measured in meters of seawater.

N

narcosis – a form of mental incapacity experienced by people while breathing an elevated partial pressure of a gas such as nitrogen.

NERD (Near Eye Rebreather Display) – a heads-up display that duplicates the wrist unit or primary controller.

GLOSSARY

nitrogen narcosis – a form of narcosis induced by breathing an elevated partial pressure of nitrogen (N2).

nitrox – see enriched air nitrox.

no-decompression dive – any dive that allows a diver to ascend directly to the surface, without the need for staged decompression stops.

normoxic – a mixture of gas containing 21 percent oxygen.

O

O_2 control system – the components of a rebreather which manage the concentration of oxygen in the breathing loop. The system usually includes sensors, electronics and a solenoid valve that injects oxygen.

N = NERD - a new style of heads-up display

O_2 sensor – a galvanic cell that generates an electrical voltage that is proportional in strength to the partial pressure of oxygen exposed to the sensor.

offboard diluent – a diluent gas tank that is clipped externally to a rebreather.

offboard oxygen – an oxygen tank that is clipped externally to a rebreather.

OLED (Organic Light-Emitting Diode) - a display type that does not use a backlight and is able to display rich blacks that offer greater contrast in low light applications such as diving.

onboard diluent – a diluent tank that is integrally mounted on a rebreather.

onboard diluent regulator – a first-stage regulator which attaches to the onboard diluent tank of a rebreather.

onboard oxygen – a oxygen tank that is integrally mounted on a rebreather.

onboard oxygen regulator – a first-stage regulator which attaches to the onboard oxygen tank.

O = Offboard Tank - A team of divers practices passing their offboard tanks to simulate emergency tank swapping.

overpressure relief valve (OPV) – The adjustable valve attached to the bottom of the exhalation counterlung, which allows excess gas and accumulated water in the breathing loop to be vented; also known as a loop vent valve.

open-circuit SCUBA (OC) – self-contained underwater breathing apparatus where the inhaled breathing gas is supplied from a high-pressure cylinder to the diver via a two-stage pressure reduction demand regulator, and the exhaled gas is vented into the surrounding water and discarded in the form of bubbles.

oxygen rebreather – a type of closed-circuit rebreather that incorporates only oxygen as a gas supply. The earliest form of closed-circuit rebreathers, designed for covert military operations, submarine escape and mine rescue operations.

GLOSSARY

oxygen toxicity – symptoms experienced by individuals suffering exposures to oxygen that are above normal ranges tolerated by human physiology.

P

partial pressure – the portion of the total gas pressure exerted by a single constituent of a gas mixture calculated by multiplying the fraction of the gas by the absolute pressure of the gas.

passive addition – gas addition systems utilized by some semi-closed circuit rebreathers to passively inject gas into the breathing loop; usually achieved by a mechanical valve that opens in response to a collapsed bellow or drop in breathing loop gas pressure.

PN_2 – the partial pressure of nitrogen in a gas mixture, usually referring specifically to the breathing gas mixture inhaled by a diver.

P = Practice - If you have been away from your rebreather for a while, schedule a practice dive.

PO_2 – the partial pressure of oxygen in a gas mixture, usually referring specifically to the breathing gas mixture inhaled by a diver.

PO_2 setpoint – the PO_2 set by the diver, used to determine when a solenoid valve injects oxygen into the breathing loop.

pulmonary oxygen toxicity – severe hyperoxia caused by prolonged exposure to breathing mixtures with partial pressures in excess of 0.5 ATA. This form of oxygen toxicity primarily affects the lungs and causes pain on deep inhalation as well as other symptoms.

R

radial CO_2 absorbent canister (radial scrubber) – a cylindrical CO_2 absorbent canister design wherein the gas flows laterally from the outside to the inside of a hollow tube (or vice-versa), like a donut.

rebreather – any form of life-support system where the user's exhaled breath is partially or entirely re-circulated for subsequent inhalation.

P = Pace - Pace yourself during dives that require exertion.

GLOSSARY

respiratory minute volume (RMV) – the volume of gas inhaled and exhaled during one minute of normal breathing.

RMV – see respiratory minute volume.

S

safety stops – any staged decompression stops prior to ascending directly to the surface, even if the decompression model being followed for the dive does not require any such stops.

scrubber – see CO_2 absorbent.

semi-closed circuit rebreather (SCR) – a type of rebreather that injects a mixture of nitrox or mixed gas into a breathing loop to replace that which is used by the diver for metabolism; excess gas is periodically vented into the surrounding water in the form of bubbles.

setpoint – see PO_2 setpoint.

R = RMV - Use your RMV to help you calculate your SAC rate so that you can plan for adequate bailout gas.

S = Scooter - Gain some experience before jumping back onto your DPV. New issues face scooter drivers such as buoyancy, monitoring gauges and drag.

shoulder port – the plastic shoulder connectors in a breathing loop which connect the breathing hoses to the counterlungs, sometimes serving as water traps to divert condensation and leaked water into the counterlungs and down to the OPV.

skip breathing - the practice of inhaling, holding the breath and then exhaling in order to attempt to extend the time underwater by using less air. This practice often leads to buildup of carbon dioxide and hypercapnia.

Sodalime – a general term referring to a chemical agent which reacts and bonds with carbon dioxide and is commonly used in the scrubbers of rebreathers.

sodasorb – see HP Sodasorb.

Sofnolime® – a brand of CO_2 absorbent material.

solenoid valve – a valve that opens when electricity is applied to an electromagnetic solenoid coil; usually the type of valve used to inject oxygen into the breathing loop of a closed-circuit rebreather.

stack – slang terminology referring to the carbon dioxide absorbent canister.

stack time – a term used to describe the predicted time that a canister of CO_2 absorbent will last before it needs to be replaced.

223

GLOSSARY

T = Tom Iliffe - Coasting through the Tunnel To Atlantis in Lanzarote, finding new species of cave adapted animal life.

T

trimix – a gas mixture containing three constituents; usually oxygen, nitrogen, and helium.

U

upstream – a relative direction with respect to the flow of gas through the breathing loop of a rebreather; the opposite of downstream.

upstream check-valve – a one-way valve system that permits inhaled gas to flow from the inhalation breathing hose to the mouthpiece, but prevents exhaled gas from flowing backwards. This valve is part of the breathing loop system that enables circular flow of gas.

V

venting breath – a type of breathing pattern used to purge gas from a breathing loop; accomplished by inhaling through the mouth and exhaling through the nose into the mask or around the edge of the mouthpiece, thus creating bubbles.

W

whole-body oxygen toxicity – see pulmonary oxygen toxicity.

work of breathing – the effort required to complete an inspiration and expiration cycle of breathing. Work of breathing can be affected by breathing hose diameters, check valve design, scrubber design, counterlung placement and design, depth, absorbent material and other factors.

U = Cenote Ucil - The author and her team dive in a deep cenote during a National Geographic project in Mexico.

workload – a representation of the level of physical exertion.

The Last Word

If you decide to take up rebreather diving, you may be preparing to sign up for a class now. You will be asked to affirm that you understand and accept the following risks as well as others. These pages act as a of a review of takeaway lessons from this book. Though they are densely worded, they should offer you time for thoughtful consideration before enrolling in class.

Diving a rebreather carries with it certain responsibilities and risks that are not present when using open circuit SCUBA equipment. Extensive experience diving with open circuit SCUBA equipment does not necessarily transfer well to diving with a rebreather. You must acquire all the knowledge, skills and training from a factory recognized instructor that are necessary to safely use a rebreather. When you sign up for a rebreather class you will be waiving important legal rights for yourself, your family, estate, heirs and assigns. You must fully inform your family of the potential for injury or death when using a rebreather.

You have the responsibility to always follow your training and adhere to the safe diving procedures taught in your certification course(s), including, but not limited to: using checklists, appropriate setpoints, pre-dive and post-dive maintenance of all components, the need for ongoing experience and unit-specific training, physical and medical fitness stipulations by the diver, and any other details that relate specifically to the use of rebreathers and their foreseeable risks. Failure to adhere to your training and safe diving practices will significantly increase the risk of suffering serious injury or death.

Problems may arise pre, post and during diving operations when using a rebreather, that when not dealt with properly may have fatal or near fatal consequences. It is therefore mandatory that you understand specifically how your rebreather works, the purpose of every component, maintenance of the components, the contingency planning necessary for the dive if problems or anything reasonably foreseen or unforeseen may arise and the operational planning necessary for a diving operation. You must get the necessary skills, experience, and training to properly be engaged in rebreather diving including, without limitation, diving involving the use of a specific rebreather and its components.

You will be the person who is ultimately and solely responsible and liable for the proper assembly, inspection, maintenance and operation of your rebreather and for assuring that the rebreather and its components are operational before beginning a dive, and your failure to do so may lead to a malfunction or failure of the rebreather, thereby causing injury or death.

You must ensure that you receive a copy of the instruction manual for the specific rebreather you will use, and you must read and fully understand the contents of such manual prior to attempting any dive using the rebreather, and you shall heed all cautions, warnings and instructions contained in the manual.

You must not attempt any dive using a rebreather unless and until you: (i) have been certified by a duly qualified diving instructor who is specifically qualified to certify divers in respect of the use of the specific rebreather; or are undergoing training for the purposes of obtaining such certification; (ii) have been specifically trained and certified by a duly qualified instructor in the technology of using and mixing oxygen and other gases, and have obtained and been trained in the use of all equipment required for the handling and mixing of gases including, without limitation, those used in connection with the rebreather; (iii) have independently ensured that all gases and chemicals being used have been properly prepared prior to their use; and (iv) have set up and assembled the rebreather using the procedures set forth in the instruction manual for the rebreather, and you have double-checked all pre-dive, dive and post-dive procedures using appropriate checklists.

Before diving a rebreather, you must be fully aware that there are many inherent risks and hazards associated with SCUBA diving and using a rebreather. No piece of equipment is foolproof, and while diving a rebreather you will or may be exposed to a number of hazards, including, but not limited to: (a) the risks of equipment malfunction or failure, including that which may result from the design, assembly, or manufacturer of a rebreather; (c) risk of injury caused by lack of or inadequate instruction or warning; (c) risks arising from improper and/or negligent operation of a rebreather; (d) inherent risks and hazards related to diving in an underwater environment, including, but not limited to: decompression sickness, barotrauma, arterial gas embolism, hyperoxia, hypoxia, hypercapnia, narcosis, other hyperbaric injuries, drowning and other hazards relating to SCUBA diving, and death. You must fully understand and accept that, by participating in SCUBA diving and using a rebreather, you may be seriously injured or killed, even if you do nothing wrong.

You must understand that all electronic controls, instruments and breathing loops associated with rebreathers are inherently subject to random and spontaneous failure, and that those subjected to aquatic environments are especially prone to failure. As such, you must agree and acknowledge that you will not conduct any dive operations using a rebreather without also carrying the appropriate redundant breathing gas supply and decompression instruments/tables that you can use in the event of failure of the breathing loop or your chosen decompression system, or both. You understand that you must continuously monitor a rebreathers' electronic controllers to watch for warnings and/or failures in the system, and you must agree to do so. You further understand that the sole responsibility for keeping you alive underwater rests with you and not the equipment, and you must agree to plan and conduct all dive operations so they may be conducted safely and successfully.

You must understand that it takes time to become fully familiar with using a rebreather for diving, and it takes time diving with a rebreather to become proficient with the apparatus and operational procedures. You must understand that experience comes with time and will be built up gradually, and you will dive the rebreather on a reasonable and consistent basis to maintain proficiency in its use, including practicing buoyancy control. You must fully understand that, as part of the operational planning of diving a rebreather, you will properly plan the diving operation to cover all those reasonably foreseeable contingencies that may appear during the diving operation, especially complete failure of the rebreather, and that you will be equipped to safeguard yourself and those that you are diving with.

You must understand that any modifications or alterations to a rebreather of any kind, other than those made by the manufacturer, are extremely dangerous, strictly prohibited and may void any warranty.

You rebreather is for your personal use only. You must not sell a rebreather to any person who is not properly trained and certified to use the specific rebreather, nor should you allow any person who is not properly trained and certified to use the specific rebreather.

Though intensely legalistic, these words (borrowed from liability release documents of various manufacturers) are offered to give you the opportunity to fully consider and accept the risks of rebreather diving. Big Sister Jill wants you to have as many tools as possible available to make a good decision for yourself and your family. With the chance to review these considerations now, the purchase and classroom paperwork that lays ahead won't be too much of a surprise!

Now, be safe and have fun!

Cave Diver Jeff Wollenberg glides through the tunnels of Ginnie Springs.

www.ingramcontent.com/pod-product-compliance
Lightning Source LLC
Chambersburg PA
CBHW041117300426
44112CB00002B/10